Lecture Notes in Mathematics

Edited by A. Dold and B. Eckmann

615

Turbulence Seminar

Berkeley 1976/77

Organized by
A. Chorin, J. Marsden and S. Smale

Edited by
P. Bernard and T. Ratiu

Springer-Verlag
Berlin Heidelberg New York 1977

Editors

Peter Bernard
Institute for
Physical Sciences and Technology
University of Maryland
College Park, Maryland 20742/USA

Tudor Ratiu
Mathematics Department
University of California, Berkeley
Berkeley, CA 94720/USA

AMS Subject Classifikations (1970): 34 C 35, 76 D 05, 76 F 05

ISBN 3-540-08445-2 Springer-Verlag Berlin Heidelberg New York
ISBN 0-387-08445-2 Springer-Verlag New York Heidelberg Berlin

Printing and binding: Beltz Offsetdruck, Hemsbach/Bergstr.
2141/3140-543210

FOREWORD

A good introduction to this volume could be the verbatim announcement of the seminar from which these lectures were taken. In fact, the seminar took place as announced, the attendance ranged around 80 or 90 for each lecture, notes of the expositions were taken by Tudor Ratiu and Peter Bernard, and this volume is the result.

Alexandre Chorin
Jerrold Marsden
Steve Smale

The actual Turbulence Seminar developed threefold: a series of eight main lectures and two parallel student oriented seminars dealing with dynamical systems and numerical methods in fluid dynamics, conducted respectively by Charles Pugh and Alexandre Chorin. The most relevant talks were included in the volume as appendices and they were written up directly by the authors. The only exception is the Appendix to Lecture VII by Oscar Lanford who kindly offered the present computer pictures of the Lorenz attractor for this volume after the seminar was over. The notes of Lecture V by Benoit Mandelbrot were written by the author too. Unfortunately Lecture VI offered by W. Kline of Stanford University is not available. The note-taking is divided as follows: Lecture IV and II are written by Peter Bernard (II with the contribution of T. Ratiu) and the rest by Tudor Ratiu. We want to thank all the authors of the talks for their support and valuable assistance in writing up these notes,

without their help, this volume would never have appeared. Many thanks
also to all those in the audience, who, by the questions raised,
clarified for us many "dark spots" of the expositions.

Tudor Ratiu

DEPARTMENT OF MATHEMATICS
University of California, Berkeley

Announcement

Seminar on Turbulence: From Numerical Analysis to Strange Attractors
Fall 1976   Mondays at 4:00 p.m.
Chorin/Marsden/Smale

The aim is to discuss problems connected with recent theories and models
proposed for turbulence, the impact of dynamical systems theory on the subject
as well as numerical work, and the role of the Navier-Stokes Equations. The
seminar will give background on these subjects, specific examples will be
studied. Also the content will include Hopf Bifurcation, Ruelle-Takens, and
ergodic theory of dynamical systems.

We hope the seminar could give focus to serious attacks on the fundamental
problem of finding a feasible model of turbulence. After the seminar, discussions
will be continued over beer.

The following references contain some background material.

1. R. Bowen, "Equilibrium States and the Ergodic Theory of Anosov Diffeomorphisms
    Springer Lecture Notes #470 (1975).

2. A. Chorin, "Lectures on Turbulence Theory", Publish/Perish (1975).

3. L. D. Landau and E. M. Lifshitz, "Fluid Mechanics", Addison Wesley (1959).

4. J. Marsden, "A Short Course in Fluid Mechanics", Publish/Perish (1976).

5. J. Marsden and M. McCracken, "The Hopf Bifurcation and Its Applications",
    Springer Notes in Applied Math (1976).

6. P. G. Saffman, "Lectures on Homogeneous Turbulence", Topics in Nonlinear
    Physics (ed. Zabusky) Springer 1968.

7. S. Smale, "Differentiable Dynamical Systems", Bull. A.M.S. 73 (1967)
    747-817.

TABLE OF CONTENTS

LECTURE I

ATTEMPTS TO RELATE THE NAVIER-STOKES
EQUATIONS TO TURBULENCE

Jerry Marsden

The present talk is designed as a survey, is slanted
to my personal tastes, but I hope it is still represent-
ative. My intention is to keep the whole discussion pretty
elementary by touching large numbers of topics and avoiding
details as well as technical difficulties in any one of
them. Subsequent talks will go deeper into some of the
subjects we discuss today.

We start with the law of motion of an incompressible
viscous fluid. This is given by the Navier-Stokes Equations

$$\begin{cases} \dfrac{\partial v}{\partial t} - \nu \Delta v - (v \cdot \nabla)v = -\nabla p + f \\[2mm] \text{div } v = 0 \\[2mm] v = \begin{cases} 0 \quad \text{or} \\ \text{prescribed} \end{cases} \quad \text{on} \quad \partial\Omega \end{cases}$$

where Ω is a region containing the fluid, v the velocity
field of the fluid, p the pressure and f the external
forces. v represents here the kinematic viscosity, or, in
the way we wrote our equations $1/Re$, where Re is the
Reynolds number. The derivation of these equations can be
found in any book on hydrodynamics, such as Landau and
Lifschitz [1], K. O. Friedrichs and R. von Mises [1], and
Hughes and Marsden [1]. We note here that the relevance
of the incompressibility condition $\text{div } v = 0$ for turbulence is a matter for debate, but the general agreement
today seems to be that compressible phenomena are not a
necessary factor in turbulence; they start to be necessary
only at very high speeds of the fluid.

Turbulence is the chaotic motion of a fluid. Our goal
in this talk is to try to relate this universally accepted
physical definition to the dynamics of the Navier-Stokes
equations. There have been at least three attempts to
explain the nature of turbulence, each attempt offering a
model which will be briefly discussed below:

(a) The Leray picture (1934). Since the existence
theorems for the solutions of the Navier-Stokes equations
in three dimensions give only local semiflows (i.e.,
existence and uniqueness only for small intervals of time),
this picture assumes that turbulence corresponds to a breakdown of the equations after a certain interval of time; in
other words, one assumes that the time of existence of the

solutions is <u>really</u> finite. Schaffer [1] looked at those t for which the equations break down and found that this set is of Hausdorff measure $\leq 1/2$. It is hard to imagine realistic physical situations for which the Navier-Stokes equations break down.

(b) <u>The E. Hopf-Landau-Lifschitz picture</u>. This is extensively discussed in Landau-Lifschitz [1] and consists of the idea that the solutions exist even for large t, but that they become quasi-periodic. Loosely speaking, this means that as time goes by, the solutions pick up more and more secondary oscillations so that their form becomes, eventually,

$$v(t) = f(\omega_1 t, \ldots, \omega_k t)$$

with the frequencies irrationally related. For k big, such a solution is supposed to be so complicated that it gives rise to chaotic movement of the fluid.

(c) <u>The Ruelle-Takens picture</u> (1971) assumes that the dynamics are inherently chaotic.

In the usual engineering point of view, the "nature" of turbulence is not speculated upon, but rather its statistical or random nature is merely assumed and studied.

Having this picture, a main goal would be to link up the statistics, entropy, correlation functions, etc., in the engineering side with a "nice" mathematical model of turbulence. More than that, such a model must be born out of the Navier-Stokes equations. Note that in this model we believe, but do not assume, that the solutions of the Navier-Stokes equations exist for large t and that the information on the chaoticness of the fluid motion is already in the flow. Needless to say, today we are very far away from this goal. This last picture is interesting and has some experimental support (J. P. Gollub, H. L. Swinney, R. Fenstermacher [1], [2]) which seems to contradict the Landau picture. There are "nice" mathematical models intrinsically chaotic strongly related to the Navier-Stokes equations. These are the Lorentz equations obtained as a truncation of the Navier-Stokes equations for the Benard problem and whose dynamics are chaotic.

The rest of the talk is devoted to a survey of the pros and cons of these models. All the details on these will be made by means of a series of remarks.

Remark 1. In two dimensions the Navier-Stokes equations and also the Euler equations (set $\nu = 0$ in the Navier-Stokes equations, which corresponds to a non-viscous fluid) have global t-solutions. Hence, the Leray picture cannot happen in two dimensions! (Leray [1], Wolibner [1], Kato [1], Judovich [1]).

In three dimensions, the problem is open. There are no theorems and no counterexamples. However, there is some very inconclusive numerical evidence which indicates that

(a) for many turbulent or near turbulent flows, the
Navier-Stokes equations do not break down.

(b) for the Euler equations with specific initial
data on $\mathbb{T}^3$ (the Taylor - Green vortex):

$$
\begin{cases}
v_1 = \cos x \sin y \sin z \\
v_2 = -\sin x \cos y \sin z \\
v_3 = 0
\end{cases}
$$

the equations might break down after a finite time. Specifically, after $T \approx 3$, the algorithm used breaks down. This
may be due to truncation errors or to the actual equations
breaking down, quite probably the former. We only mention
that this whole analysis requires the examination of convergence of the algorithms as well as their relation to the
exact equations; see the numerical studies of Chorin [1,2],
Orszag [1] and Herring, Orszag, Kraichnan and Fox [1], Chorin
etal [1], and references therein.

Remark 2. The Landau picture predicts Gaussian statistics. This is not verified in practice. The model with
chaotic dynamics does not predict such a statistic (see
Ruelle [2], Gollub and Swinney [1]).

Remark 3. The Landau picture is unstable with respect
to small perturbations of the equations. The Ruelle-Takens

picture is, in some sense, a stabilization of the Hopf-Landau-Lifschitz picture. However, as Arnold has pointed out, strange attractors may form a <u>small</u> open set and still the quasi-periodic motions may be observed with higher probability.

<u>Remark 4</u>. Chaotic dynamics is not necessarily born from complicated equations. The Navier-Stokes equations are complicated enough to give rise to very complicated dynamics, eventually leading to a chaotic flow. The reason for this is that simple ordinary differential equations lead to chaotic dynamics (see below) and "any" bifurcation theorem for ordinary differential equations can work for Navier-Stokes equations, cf. Marsden-McCracken [1]. We do not want to go into the details here of this statement and we merely say that we look at the Navier-Stokes equations as giving rise to a vector field on a certain function space, we prove the local smoothness of the semi-flow and verify all conditions required for a bifurcation theorem; in this way we are able to discuss how a fixed point of this vector field splits into two other fixed points, or a closed orbit, and discuss via a certain algorithm their stability. Later talks with clarify and give exact statements of the theorems involved; we have in mind here the Hopf bifurcation theorem and its extension to semi-flows (see Marsden [2], Marsden and McCracken [1] and the appendix following).

<u>Remark 5</u>. As we mentioned earlier, the global t- existence theorem for the solutions of the Navier-Stokes

equations is completely open in three dimensions. It is
not necessary in the Ruelle-Takens picture of turbulence
to assume this global t-existence. If one gets an
attractor which is bounded, global t-solutions will follow.

Remark 6. There are other "simpler" partial differential equations where complex bifurcations have been
classified:

(a) Chow, Hale, Malet-Paret [1] discuss the von
Karmen equations. (This seems to be a highly nontrivial
application of ideas of catastrophe theory.)

(b) P. Holmes [1] fits the bifurcation problem for a flutter-
ing pipe into Taken's normal form.

Remark 7. There are at least two physically interacting real mathematical models with chaotic dynamics:

(a) Lorentz equations

$$\begin{cases} \dot{x} = -\sigma x + \sigma y \\ \dot{y} = rx - y - xz \\ \dot{z} = -bz + xy \end{cases}$$

(Note the symmetry
$x \mapsto -x$,
$y \mapsto -y$,
$z \mapsto z$.)

They represent a modal truncation of the Navier-Stokes
equations in the Benard problem. It is customary to set
$\sigma = 10$, $b = 8/3$; r is a parameter and represents the
Rayleigh number. We shall come back to these equations

in Remark 9.

(b) Rikitake dynamo. This model consists of two dynamos which are both viewed as generators, and as motors in interaction; it is a model for the Earth's magnetohydro-dynamic dynamo. It has also chaotic dynamics. See Cook and Roberts [1]. The equations are:

$$\dot{x} = -\mu x + zy$$

$$\dot{y} = -\mu y - \alpha x + xz$$

$$\dot{z} = 1 - xy \qquad .$$

(c) A model of mixing salt with fresh water in the presence of temperature gradients. This was communicated to me personally by H. Huppert at Cambridge.

Remark 8. In many cases, existence of center manifolds of dimension k justify a modal or other truncation to give a k-dimensional system, i.e., all the complexity really takes place in a finite dimensional invariant manifold. (Exact statements will be given in one of the next talks.)

Remark 9. For the actual Navier-Stokes equations we do not know any solutions which are turbulent, or even that they exist. In any specific turbulent flow we don't know what the chaotic attractor might look like, or how one might form. However, we do know how this works (or think we do)

for the Lorenz model. It is true that there are many
objections to my drawing conclusions about the turbulence
stemming from the Navier-Stokes equations by working with
a truncation; it is argued that truncation throws turbulence away, too. However, I think that the model of
Lorenz equations, though a truncation, can give some
insight on what may happen in the much more complicated
situation of the Navier-Stokes equations. I want to present here briefly the bifurcation for the Lorenz model
when r (the Rayleigh number) varies. The picture presented
below is due to J. Yorke, J. Guckenheimer, and O. Lanford. I am
indebted to them and to N. Kopell for explaining the results.
(See Kaplan and Yorke [1] and Guckenheimer's article in Marsden
and McCracken [1] as well as William's lecture below.)

$r < 1$: Then the origin is a global sink:

(all eigenvalues are
real and negative for
$1 > r > (4\sigma-(\sigma+1)^2/4\sigma$
i.e. $1 > r > -2.025$).

$r = 1$ and $1+\varepsilon$: At this value the first bifurcation

occurs. One real eigenvalue for the linearization at zero
crosses the imaginary axis travelling at nonzero speed on
the real axis, for the origin a fixed point. Two stable fixed
points branch off. They are at $(\pm\sqrt{b(r-1)}, \pm\sqrt{b(r-1)}, r-1)$.

This is a standard
and elementary bifur-
cation resulting in a
loss of stability by
the origin.

As r increases the two stable fixed points develop two
complex conjugate and one negative real eigenvalues. The
picture now looks like (z-axis is oriented upwards and the
plane is the x 0 z plane):

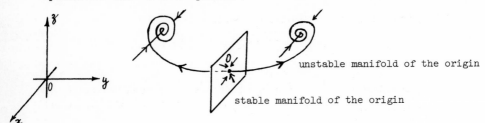

As r increases, the "snails" become more and more
inflated.

r ≅ 13.926: At around this value (found only by numerical
methods) the "snails" are so big that they will enter
the stable manifold of the origin. Stable and unstable
manifold become identical; the origin is a homoclinic
point. Another bifurcation now takes place. The
picture is, looking in along the x-axis.

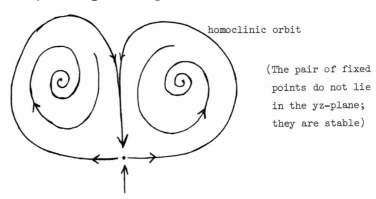

(The pair of fixed
points do not lie
in the yz-plane;
they are stable)

<u>r > 13.926</u>: The two orbits with infinite period "starting" and "ending" in the origin "cross over". The "snails" still inflate and by doing this, the homoclinic orbits leave behind unstable closed periodic orbits. The picture of the right hand side is:

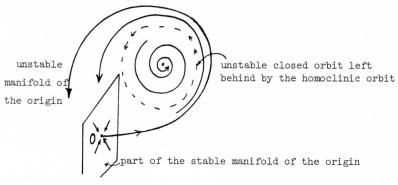

unstable manifold of the origin

unstable closed orbit left behind by the homoclinic orbit

part of the stable manifold of the origin

The unstable manifold of the origin gets attracted to the opposite fixed point for these values of r.

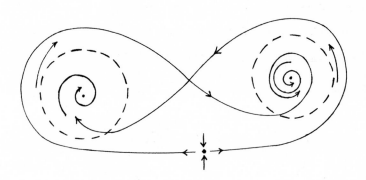

At this stage, which Yorke calls "preturbulent," there
is a horseshoe strung out between the attracting fixed points.
There are infinitely many periodic orbits, but eventually most
orbits go to one of the attracting fixed points. There is no
strange attractor, but rather a "meta-stable" invariant set;
points near it eventually leave it in a sort of probabilistic
way to one of the attracting fixed points.

To study this situation, one looks at the plane z = r-1
and the Poincaré, or once return map φ for the plane. On
this plane one draws L, the stable manifold of the origin
intersected with the plane.

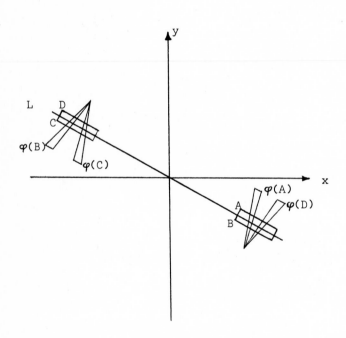

The images of the four regions A, B, C, D are shown.
If one compares this picture with Smale's horseshoe example
(Smale [1]) one sees that a horseshoe must be present. As
r increases, eventually the images of the rectangles above
will be inside themselves and an attractor will be born.
This is the bifurcation to the Lorenz attractor. Viewing
the dynamical system as a whole, we see the following (only
one half is drawn for clarity).

r = 24.06:

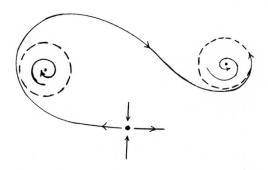

r > 24.06:

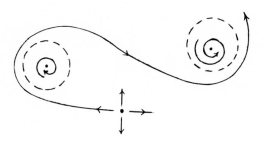

Now, between the two periodic orbits a "strange" attractor, called <u>Lorenz</u> <u>attractor</u>, is appearing. This attractor traps all the orbits that cross over the small piece of the stable manifold of the origin and throws them on the other side. Imagine we put a plane somewhere not far away from the origin, perpendicular to the drawn stable manifold and we would like to find out the points through which a specific orbit is going, travelling from one unstable closed orbit to another, and repelled by these each time; the result would be a random distribution of points in this "transveral cut" through the Lorenz attractor. For the nature of this attractor, see the talk of R. Williams in these notes, and the paper by J. Guckenheimer forming Section 12 of Marsden-McCracken [1]. We note that this attractor is nonstandard since it has two fixed points replaced by closed orbits in the "standard" Lorenz attractor. As r increases, this nonstandard Lorenz attractor grows from its initial shape and the unstable closed orbits shrink.

$r \approx 24.74 = \dfrac{\sigma(\sigma+b+3)}{(\sigma-b-1)}$: It is proved (Marsden and McCracken [1]) that a subcritical Hopf bifurcation occurs. The two closed "ghost" orbits shrink down to the fixed points which become in this way unstable.

$r > 24.74$: We now have a "standard" Lorenz attractor. The picture is:

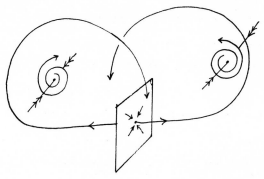

$r \gtrsim 50$. The situation for larger r is somewhat complicated and not totally settled. According to some calculations of Lanford, the following seems to happen. If we look at the once return map φ on the plane $z = r-1$, as above, then the unstable manifold of the two symmetrical fixed points develop a fold. See the following figure. When this happens, stable large amplitude closed orbits seem to bifurcate off. This folding is probably because these two fixed points are becoming stronger repellers and tend to push away the other unstable manifold.

L = stable manifold
   of the origin

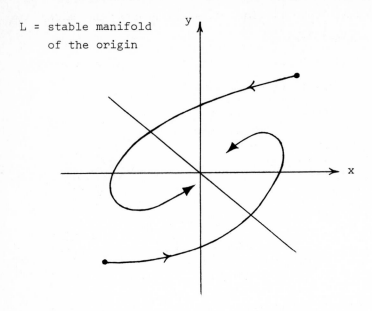

The situation is analogous to the bifurcations for
the map y = ax(1-x) which occurs in population dynamics.

One can, of course vary the other parameters in the
Lorenz model, or vary more than one. For example, Lorenz
himself in recent numerical work has looked at bifurcations
for small b (which is supposed to resemble large r).

Research projects: 1) Figure out the qualitative dynamics
and bifurcation of the Rikitake two-disc dynamo.[+]

    2) Real "pure" fluid models are needed; one might try
getting a model for:

    a) Couette Flow; see Coles [1] for many good remarks
        on this flow, and Stuart [1].

    b) Flow behind a cylinder:

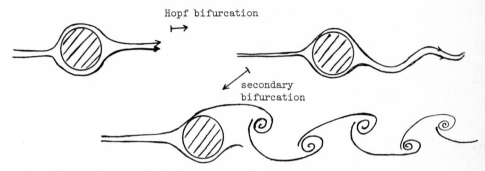

Here the symmetry will play a central role. Note that the
third picture still represents a periodic solution in the
space of divergence-free vector fields. My conjecture would
be that the secondary Hopf bifurcation is illusory and what
happens is that the original closed orbit produced by
the Hopf bifurcation gets twisted somehow in the appropriate function space.

    As A. Chorin has suggested, one should remember that the
Lorenz model is global in some sense. The choas is associated

---

[+] Some progress gas been made on this problem recently by
P. Holmes and D. Rand.

with large scale motions. One would like a model with chaotic
dynamics which is made up of a few interacting vortices and
a mechanism for vortex production. "Real turbulence" seems
to be more like this.

BIBLIOGRAPHY

CHORIN, A. J.: [1] Lectures on Turbulence Theory, nr. 5,
Publish or Perish, 1976.

CHORIN, A. J., HUGHES, T. R. J., McCRACKEN, M.J., and
MARSDEN, J. E., Product Formulas and Numerical Algorithms,
Advances in Math (to appear).

CHOW, W., HALE, J. AND MALLET-PARET, J.: [1] Generic
bifurcation theory, Archive for Rational Mechanics and
Analysis (September 1976).

COLES, D.: [1] Transition in circular Couette flow, J.
Fluid Mech. 21 (1965), 385-425.

COOK, R. and ROBERTS, J.: [1] The Rikitake two disc
dynamo system, Proc. Camb. Phil. Soc. 68 (1970), 547-
569.

FRIEDRICHS, K. O., VON MISES, R.: [1] Fluid Dynamics,
Applied Mathematical Sciences, nr. 5, Springer-Verlag,
1971.

GOLLUB, J. P., SWINNEY, H. L.: [1] Onset of turbulence in
a rotating fluid, Physical Review Letters, vol. 35,
number 14, October 1975.

GOLLUB, J. P., FENSTERMACHER, R. R., SWINNEY, H. L.:
[2] Transition to turbulence in a rotating fluid,
preprint.

HERRING, J. R., ORSZAG, S. A., KRAICHNAN, R. H., and
FOX, D. G.: J. Fluid Mech. 66 (1974), 417.

HOLMES, : [1] Bifurcation to divergence and flutter in a
pipe conveying fluid (preprint).

HOPF, E.: [1] A mathematical example displaying the features
of turbulence, Comm. Pure Appl. Math. 1 (1948), 303-322.
[2] Repeated branching through loss of stability,
an example, Proc. Conf. on Diff. Eq'ns., Maryland (1955).
[3] Remarks on the functional-analytic approach
to turbulence, Proc. Symp. Appl. Math. 13 (1962), 157-163.

HUGHES, T., MARSDEN, J.: [1] A Short Course In Fluid
Mechanics, nr. 6, Publish or Perish, 1976.

JUDOVICH, V.: [1] Mat. Sb. N. S. 64 (1964), 562-588.

KATO, T.: [1] Arch. Rat. Mech. An. 25 (1967), 188-200.

KAPLAN, J. L. and YORKE, J. A. [1] Preturbulent Behavior in the Lorenz equations (preprint).

LANDAU, L. D., LIFSCHITZ, E. M.: [1] Fluid Mechanics, Oxford: Pergamon, 1959.

LERAY, J.: [1] Sur le mouvement d'un liquide visqueux emplissant l'espace, Acta Math. 63 (1939), 193-248.

LORENZ, E. N.:[1] Deterministic nonperiodic flow, Journ. Atmospheric Sciences, 20 (1963), 130-141.

MARSDEN, J., McCRACKEN, J.: [1] The Hopf Bifurcation, Applied Mathematical Sciences 19, Springer-Verlag, 1976.

MARSDEN, J.: [2] The Hopf bifurcation for nonlinear semi-groups, BAMS, volume 79, nr. 3, May 1973, 537-541.

MARSDEN, J.: [3] Applications of Global Analysis to Mathematical Physics, Publish or Perish, 1974.

MANDELBROT, B.: [1] Geometrie fractale de la turbulence. Dimension de Hausdorff, dispersion et nature des singularites du mouvement des fluides, C. R. Aca. Sci. Paris 282 (1976), 119-120.

ORSZAG, S.A.: [1] Numberical simulation of the Taylor-Green vortex, in "Computing Methods in Applied Sciences and Engineering", Ed. R. Glowinski and J. L. Lions, Springer .1974.
    [2] Analytical theories of turbulence, J. Fluid Mech. 41 (1970), 363-386.

RUELLE, D., TAKENS, F.: [1] On the nature of turbulence, Comm. Math. Phys. 20 (1971), 167-192, and 23 (1971), 343-344.

RUELLE, D.: [2] The Lorenz attractor and the problem of turbulence, Report at the conference on "Quantum Models and Mathematics", in Bielefeld, September 1975.

SCHEFFER, V.: [1] Géométrie fractale de la turbulence. Equations de Navier-Stokes et dimension de Hausdorff, C. R. Acad. Sci., Paris (1976), 121-122.

SMALE, S.: [1] Differentiable dynamical systems, BAMS 73 (1967), 747-817.

STUART, J. T.:[1] Nonlinear Stability Theory, Annual Rev. Fluid Mech. 3 (1971) , 347-370.

WOLIBNER, W.:[1] Un théorème sur l'existence du mouvement plan d'un fluide parfait homogène, incompressible, pendant un temps infiniment longue, Math. Zeit. 37 (1933), 698-726.

APPENDIX TO LECTURE I: BIFURCATIONS,

SEMIFLOWS, AND NAVIER-STOKES EQUATIONS

Tudor Ratiu

As was pointed out in J. Marsden's talk, the Ruelle-Takens picture for turbulence assumes that the motion of the fluid is inherently chaotic, that the flow obtained for Re = 0 (solutions of the Stokes equations) gets more and more complicated as the Reynolds number Re increases, due to bifurcation phenomena until it eventually gets trapped into a "strange" attractor which has chaoticness as one of its main features. In this talk I shall summarize the mathematical results involved in this machinery, trying to back up with exact statements of theorems many exciting ideas presented in Marsden's exposition. The main source of this talk is Marsden-McCracken [1].

The leading idea is to obtain a model born out of the Navier-Stokes equations for homogeneous, incompressible, viscous fluids:

$$\begin{cases} \dfrac{\partial v}{\partial t} + (v \cdot \nabla)v - \nu \Delta v = -\text{grad } p + f \ , \quad \nu = 1/\text{Re} \\[2mm] \text{div } v = 0 \\[2mm] v = \text{prescribed on } \ \partial M \ , \text{ possibly depending on } \ \nu \quad . \end{cases}$$

Everything takes place in a compact Riemannian manifold M with
smooth boundary ∂M , v representing the velocity field of the
fluid, p the pressure and f the external force exercised on
the moving fluid. As already mentioned, Euler's equations for
an ideal fluid are obtained by setting $\nu = 0$ in the above equations; it is a theorem that the solutions to the Euler equations
are obtained as a strong limit in the H^S-topology for
$s > (\dim M)/2+1$ (see Ebin-Marsden [1]). Also notice that in
Euler's equations we have to change the boundary conditions to
$v \| \partial M$. The intuitive reason why this is so is that our fluid,
being ideal, has no friction at all on the walls; however, a
much more subtle mathematical analysis of the above described
limit process yields formally the same result, cf. Marsden [2],
Ebin-Marsden [1].

Now we would like to write our Euler and Navier-Stokes
equations in the form of a system of evolution equations

$$\frac{dv}{dt} = X_\nu(v) \ , \ v(0) = \text{given} \ ,$$

where X_ν is a densely defined nonlinear operator on a function
space picked in such a way that our boundary conditions and
$\text{div } v = 0$ should be automatically satisfied. The answer to this
question is given by the Hodge Decomposition Theorem.

Denote by $W^{s,p}$ the completion of the normed vector space
of vector-valued C^∞-functions on M under the norm

$$\|f\|_{s,p} = \sum_{0 \leq t \leq s} \|D^t f\|_{L^p} \; ;$$

here $D^t f$ denotes the differential of f , $s \geq 0$ and $1 < p < \infty$. $W^{s,p}(M)$ is the set of vector fields of class $W^{s,p}$ on M . Note that a function is of class $W^{s,p}$ if and only if all its derivatives up to order s are in L^p .

<u>Hodge Decomposition Theorem</u>. Let M be a compact Riemannian manifold with boundary and $X \in W^{s,p}(M)$, $s \geq 0$, $1 < p < \infty$. Then X has a unique decomposition

$$X = Y + \text{grad } f$$

where $\text{div } Y = 0$, $Y \| \partial M$, $Y \in W^{s,p}(M)$ and f is of class $W^{s+1,p}$.

Denote $\tilde{W}^{s,p}(M) = \{X \in W^{s,p}(M) \,|\, \text{div } X = 0, X \| \partial M\}$. Apply now the Hodge Theorem and get a map $P: W^{s,p}(M) \to \tilde{W}^{s,p}(M)$ via $X \to Y$. Let us now reformulate the Euler equations: suppose $s > n/p$; find $v: (a,b) \to \tilde{W}^{s+1,p}(M)$ such that

$$\frac{dv(t)}{dt} + P((v(t) \cdot \nabla)v(t)) = 0$$

(plus initial data). We need to assume $s > n/p$ in order to insure that the product of two elements of $W^{s,p}$ is in $W^{s,p}$ (see Adams [1], page 115). In this way, if $v \in \tilde{W}^{s+1,p}(M)$,

$(v \cdot \nabla)v \in W^{s,p}(M)$ and we can apply the Hodge Theorem. In doing this we tacitly assume that the external force is a gradient.

In order to be able to write in a similar way the Navier-Stokes equations, we change the function space to

$\tilde{W}_0^{s,p} = \{X \in W^{s,p}(M) \mid \mathrm{div}\, X = 0, X \mid \partial M = 0\}$. Then the Navier-Stokes equations can be reformulated: find $v: (a,b) \rightarrow \tilde{W}_0^{s+1,p}$ such that

$$\frac{dv(t)}{dt} - \nu P(\Delta v(t)) + P((v(t) \cdot \nabla)v(t)) = 0 \quad .$$

The following theorem is proved in Section 9 of Marsden-McCracken.

Theorem. *The Navier-Stokes equations in dimensions* 2 *or* 3 *define a smooth local semiflow on* $\tilde{W}_0^{s,2}$, *i.e., we have a collection of maps* $\{F_t^\nu\}$ *for* $t \geq 0$ *satisfying:*

(a) F_t^ν *is defined on an open subset of* $[0,\infty) \times \tilde{W}_0^{s,2}$;

(b) $F_{t+s}^\nu = F_t^\nu \circ F_s^\nu$;

(c) F_t^ν *is separately (hence, jointly)[*] continuous;*

(d) *for each fixed* t, ν , F_t^ν *is a* C^∞-*map, i.e.,* $\{F_t^\nu\}$ *is a* smooth semigroup. *More, our semiflow* $\{F_t^\nu\}$ *satisfies the so called* continuation assumption, *namely, if* $F_t(x)$ *lies in a bounded set of* $\tilde{W}_0^{s,2}$ *for each fixed* x *and for all* t *for which* $F_t^\nu(x)$ *is defined, then* $F_t(x)$ *is defined for all* $t \geq 0$.

Also, $F_t^\nu(x)$ *is jointly smooth in* t, x, ν *for* $t > 0$

[*] See Chernoff-Marsden [1], Chapter 3, or Marsden-McCracken [1], Section 8A, for the proof of the fact that separate continuity ⇒ joint continuity.

This result which goes back to Ladyzhenskaya [1] encourages us to not work with the Navier-Stokes equations under their classical form, but rather with the evolution equations in $\tilde{W}_0^{s,2}$ which they define and to analyze more closely their semiflow which has such pleasant properties.

Following the idea of chaotic dynamics, we may try to show that turbulence occurs after successive bifurcations of the solutions of the Navier-Stokes equations. Hence a first question is how much of the classical bifurcation theory can be obtained for semiflows. The work of Marsden shows that almost everything works, if one mimics the conditions on the semiflow from those, one usually has for vector fields. We shall summarize these results below.

Hence we have to cope with a system of evolution equations of the general form

$$\frac{dx}{dt} = X_\mu(x) \ , \ x(0) = given \ ,$$

where X_μ is a nonlinear densely defined operator on an appropriate Banach space E , usually -- as we already saw -- a function space and μ is a parameter. We assume that our system defines unique local solutions generating a semiflow F_t^μ for $t \geq 0$. The assumptions made on the semiflow are (a), (b), (c) and (d) above. We also ask for the continuation assumption described before. It may seem that we force our assumptions on the semiflow such as to suit our particular problem. In reality it is exactly

the other way around: one usually has these conditions satisfied
and checks them for the Navier-Stokes equations -- and this is
hard work involving a serious mathematical machinery (see Section
9 of Marsden-McCracken). It is true that the continuation
assumption might seem strong; but it merely says that we have
at our disposal a "good" local existence theorem, so "good" as
to insure the fact that an orbit fails to be defined only if it
tends to infinity in a finite time. That makes sense physically,
looking at expected solutions of the governing equations of the
law of motion of a fluid (Navier-Stokes): a solution fails to
exist only if it "blows up". Another remark is of mathematical
character and concerns the generator X_μ ; this is not a smooth
map from E to E , hence we cannot expect smoothness of
$F_t^\nu(x)$ in t . The fact is that the trouble is actually only at
t = 0 , as can be seen from the theorem on the Navier-Stokes
semiflow from before, and exactly the derivative at t = 0
gives the generator. The next group of assumptions regards the spectrum
of the linearized semiflow relevant for the Hopf bifurcation.
Spectrum Hypotheses. Let $F_t^\mu(x)$ be jointly continuous in
t,μ,x for t > 0 and μ in an interval around $0 \in \mathbb{R}$.
Suppose in addition that:

  (i)   0 is a fixed point of F_t^μ , i.e., $F_t^\mu(0) = 0$, $\forall \mu, t$;

  (ii) for μ < 0 , the spectrum of $G_t^\mu = DF_t^\mu(0)$ is contained
        inside the unit disc $D = \{z \in C \mid |z| < 1\}$;

  (iii) for μ = 0 (resp. μ < 0) the spectrum of G_1^μ at the
        origin has two isolated simple eigenvalues λ(μ) and

$\overline{\lambda(\mu)}$ with $\lambda(\mu) = 1$ (resp. $\lambda(\mu) > 1$) and the rest of the spectrum is in D and remains bounded away from the unit circle;

(iv) $\left.\dfrac{d|\lambda(\mu)|}{dt}\right|_{\mu=0} > 0$, i.e., the eigenvalues move steadily across the unit circle.

Sometimes we look at these hypotheses but with (iii) changed to:

(iii') for $\mu = 0$ (resp. $\mu < 0$) the spectrum of G_1^{μ} at the origin has one isolated simple real eigenvalue $\lambda(\mu) = 1$ (resp. $\lambda(\mu) > 1$) and the rest of the spectrum is in D and remains bounded away from the unit circle;

(v) for $\mu = 0$ the origin is asymptotically stable.

We won't go into the technical details of this last hypothesis here and say only that it involves an algorithm of checking if a certain displacement function obtained via Poincaré map has strictly negative third derivative.

Bifurcation to Periodic Orbits: *Under the above hypotheses (i)-(v) there is a fixed neighborhood* V *of* 0 *in* E *and an* ε > 0 *such that* $F_t^{\mu}(x)$ *is defined for all* t ≥ 0 *for* μ ∈ [-ε,ε] *and* x ∈ V . *There is a one-parameter family of closed orbits for* F_t^{μ} *for* μ > 0 , *one for each* μ > 0 *varying continuously with* μ . *They are locally attracting and*

hence stable. Solutions near them are defined for all $t \geq 0$.
There is a neighborhood U *of the origin such that any closed
orbit in* U *is one of the above orbits.*

Bifurcation to Fixed Points: *Same hypothesis with (iii) and
(iii') interchanged. Then the same result holds, replacing
the words "closed orbit" with "two fixed points".*

I shall not go into the proof of these theorems but will
give the two crucial facts behind the formal proof. One is the
Center Manifold Theorem and the other is a theorem of Chernoff-
Marsden regarding smooth semiflows on finite-dimensional manifolds. Coupling these two results reduces the whole problem to
the classical Hopf Bifurcation Theorem in 2 dimensions, which
is relatively simple and goes back to Poincaré. Here are the
statements:

Center Manifold Theorem for Semiflows: *Let* Z *be a Banach
space admitting a* C^∞-*norm away from zero, and let* F_t *be a
continuous semiflow defined in a neighborhood of zero for*
$0 \leq t \leq z$. *Assume* $F_t(0) = 0$ *and that for* $t > 0$, $F_t(x)$
is jointly C^{k+1} *in* t *and* x . *Assume that the spectrum of
the linear semigroup* $DF_t(0): Z \to Z$ *is of the form* $e^{t(\sigma_1 \cup \sigma_2)}$
where $e^{t\sigma_1}$ *lies on the unit circle (i.e.,* σ_1 *lies on the
imaginary axis) and* $e^{t\sigma_2}$ *lies in the unit circle at non-zero
distance from it for* $t > 0$ *(i.e.,* σ_2 *is in the left half*

plane). Let Y be the generalized eigenspace corresponding to the spectrum on the unit circle; assume $\dim Y = d < +\infty$. *Then there exists a neighborhood of* 0 *in* Z *and a* C^k-*submanifold* $M \subseteq V$ *of dimension* d *passing through* 0 *and tangent to* Y *at* 0 *such that:*

(a) *Local Invariance: if* $x \in M$, $t > 0$ *and* $F_t(x) \in V$, *then* $F_t(x) \in M$;

(b) *Local Attractivity: if* $t > 0$ *and* $F_t^n(x)$ *remains defined and in* V *for all* $n = 0,1,2,\ldots$, *then* $F_t^n(x) \to M$ *as* $n \to \infty$.

This is applied to F_t^μ after suspending μ to obtain the semi-flow $F_t(x,\mu) = (F_t^\mu(x),\mu)$ on the original space x the parameter space.

The version of this theorem for a C^{k+1} map is well known; however, this statement regarding semiflows -- although believable -- wasn't present in the literature before; the first time it appears is in Section 2 of Marsden-McCracken. Note that every-thing works out nicely in the theorem, even though the generator X of the semiflow is unbounded.

<u>Theorem (Chernoff-Marsden)</u>: *Let* F_t *be a local semiflow on a Banach manifold* N *jointly continuous and* C^k *in* $x \in N$. *Suppose that* F_t *leaves invariant a finite dimensional submanifolf* $M \subseteq N$. *Then on* M, F_t *is locally reversible, is jointly* C^k *in* t *and* x *and is generated by a* C^{k-1} *vector field on* M.

Some remarks are in order. Besides being one key factor in the proof of the bifurcation theorem, the center manifold theorem might justify some modal truncations of the Navier-Stokes

equations to give a d-dimensional system (see Remark 8 of
Lecture I by J. Marsden). Also, in Marsden-McCracken, Section
4A, an algorithm is described which enables us to check on the
stability of the new born fixed points or closed orbits after
bifuracations. Remark 4 of Lecture I hints toward that. The
reduction to two dimensions appears as a corollary of the
proof of the Bifurcation Theorem. The conclusion is that all
the complexity in this case takes place only in a plane, even
though we started off with an evolution equation on an infinite
dimensional function space. This occurrence is characteristic
when we work with semiflows; trying to prove a bifurcation,
we reduce everything to a finite dimensional theorem for flows
and this gives us then two things: the theorem itself and the
reduction!

That's the way one approaches the next bifurcation to
invariant tori. Here the Hopf Bifurcation Theorem for Diffeomorphisms will be needed and the idea of the proof is the same
as before; one has to replace the argument of the Hopf Bifurcation Theorem in $\mathbb{R}^2$ with a similar argument using now the Hopf
Bifurcation Theorem for Diffeomorphisms. I won't go into any
technical details.

That would roughly solve the approach to the first two
bifurcations. How about higher ones? The only leading idea
is the Poincaré map, and the fact that something invariant for
it, yields an invariant manifold of one higher dimension for

the semiflow with the preservation of the attracting or repelling character: a fixed point -- attracting or repelling --
gave a closed orbit -- attracting or repelling -- a circle,
an invariant torus, etc.

Let me mention that all these geometrical methods presented
here are by no means the only ones with which one could attack
bifurcation problems for the Navier-Stokes equations. An excellent reference is J. Sattinger [1], who in Chapters 4-7 does
roughly the same thing, but using methods of eigenvalue problems,
energy methods and Leray-Schauder degree theory. I prefer the
above methods because I think they appeal more to one's
geometrical intuition.

As a concluding remark, let me say that even if it seems
that the first bifurcations can be attacked successfully with
the above methods, the difficulties one faces might be very
big. One has to start off with something known, namely a
particular stationary solution, regard this as a fixed point
of the generator of the semiflow and work his way through the
conditions in the Bifurcation Theorem. In many cases we do
not have even a stationary solution! In the research problem
suggested in Lecture I about the flow behind a cylinder, the
difficulty is exactly this one: there is no explicitly solution known (for Re > 0) of the laminar flow

in 2 or 3 dimensions, let alone of more complicated situations.

BIBLIOGRAPHY

ADAMS, R.: [1] Sobolev Spaces, Academic Press, 1975, in the
    Series of Pure and Applied Mathematics, volume 65.

CHERNOFF, P., MARSDEN, J.: [1] Properties of Infinite Dimen-
    sional Hamiltonian Systems, Springer Lecture Notes in
    Mathematics, volume 425, 1974.

EBIN, D., MARSDEN, J.: [1] Groups of diffeomorphisms and the
    motion of an incompressible fluid, Ann. of Math., volume
    92, no. 1, July 1970, 102-163.

HUGHES, T., MARSDEN, J.: [1] A Short Course in Fluid Mechanics,
    Publish or Perish, 1976.

LADYZHENSKAYA, O.: [1] The Mathematical Theory of Viscous
    Incompressible Flow, Gordon and Breach, N.Y., 1969.

MARSDEN, J.: [1] The Hopf Bifurcation for nonlinear semigroups,
    BAMS, volume 76, no. 3, May 1973, 537-541.

MARSDEN, J., McCRACKEN, M.: [1] The Hopf Birfurcation, Applied
    Mathematical Sciences 19, Springer Verlag, 1976.

MORREY, C. B.: [1] Multiple Integrals in the Calculus of
    Variations, Springer, 1966.

SATTINGER, J.: [1] Topics in Stability and Bifurcation Theory,
    Springer Lecture Notes in Mathematics, volume 309, 1973.

RUELLE, D., TAKENS, F.: [1] On the nature of turbulence,
    Comm. Math. Phys. 20 (1971), 167-192.

LECTURE II

THEORIES OF TURBULENCE

Alexandre Chorin

An important reason for studying the qualitative features
of turbulence using the methods of pure mathematics is that in
this way justifications can be found for the statistical pro-
cedures that engineers use to solve actual turbulence problems.
The practice of assuming that a mean velocity field exists
whose evolution is governed by equations obtained by averaging
the Navier-Stokes equations needs rigorous proof that it leads
to a well defined problem for which a solution does exist.

The preoccupation of earlier analytical theories of turbulence with the possible breakdown of the Navier-Stokes equations
as the cause of turbulence is, from a physicist's point of view,
unwarranted. The flow of a fluid in turbulent conditions satisfies very well the hypotheses used in deriving the Navier-
Stokes equations from Newton's laws. In particular, the size of
the smallest eddies appearing in a turbulent flow is at least
three orders of magnitude larger than that of the mean free path

for a fluid at all except very extreme conditions.

The idea of Hopf and Landau that turbulence could be represented as a quasi-periodic solution of the Navier-Stokes equations is unfounded, because the flows that would result have properties which are incompatible with the properties of real turbulence.

Though the earlier analytical treatments of turbulence were off the mark, recent work in dynamical systems concerning the nature of turbulence apparently does correspond qualitatively with what one sees in the real world. In some real flows we see bifurcations and then turbulence, i.e., something with the properties of a strange attractor. The dynamical systems approach has a long distance to travel until the models it studies truly mirror the properties of real turbulent flows. For example, the problem defined by Lorenz is a model for low Rayleigh number convection, which does not display an energy cascade into high wave numbers because the small scale motion is damped by gravity. Thus, a "turbulence" with only one scale of motion may be taking place. Real turbulence is characterized by qualitatively different types of motion at a number of different scales. The Rikitake dynamo is similarly only slightly related to typical turbulence since it comes from a problem in which the set of stationary solutions of the equations is dense in the set of all solutions. Furthermore, the effects of truncation are unclear but major. The limitations of the dynamical systems approach, however, are not just in the lack of a sophisticated enough model which displays most of the features of turbulent flow. One also cannot expect to obtain quantitative results for a particular flow using these methods.

Now let us look into the quantitative side of turbulence. How are we to obtain useful information of a practical sort about a turbulent flow? The Navier-Stokes equations are:

$$\underline{u}_t + (\underline{u} \cdot \underline{\nabla})\underline{u} = -\frac{1}{\rho} \nabla p + \nu \nabla^2 \underline{u}$$

$$\nabla \underline{u} = 0 \quad ,$$

governing the motion of an incompressible fluid moving with velocity $\underline{u}$, and contained in a region D with boundary ∂D . When the viscosity ν is large, we have, in general, a laminar flow, i.e., a smooth orderly flow. When ν becomes small enough the flow becomes wildly chaotic and it is then beyond our ability to follow its detailed motion. We then seek to compute average properties of the flow field.

Equations for an average velocity $\underline{U}$ may be obtained by writing the velocity field as $\underline{u} = \underline{U} + \underline{u}'$ where $\overline{\underline{u}'} = 0$ (the overbar denotes an average), substituting this into the Navier-Stokes equations and then averaging. This results is e.g., for the momentum equation:

$$\frac{\partial U_i}{\partial t} + (\underline{U} \cdot \underline{\nabla})U_i + \overline{\nabla(u_i' \underline{u}')} = -\frac{1}{\rho}\frac{\partial p}{\partial x_i} + \nu \nabla^2 U_i \ .$$

These equations are formally identical to the Navier-Stokes equations, except for the additional term containing the quadratic velocity moments $\overline{u_i' u_j'}$. These averages are additional unknowns in the equations of motion. Equations may be obtained governing their evolution by multiplying the Navier-Stokes equations by $\underline{u}'$ and averaging. This introduces third degree products of the velocity field which require more equations for their development. If this process is continued indefinitely, it results in an infinite number of equations for an infinite number of unknowns. We clearly cannot solve such a problem unless some sort of closure is formulated which will truncate the hierarchy of equations and unknowns, by relating at some point the mean of the product of $N + 1$ velocities to the lower order products. One assumption which has been tried was to suppose that the fourth moments of $\underline{u}'$ are related to the second moments as if $\underline{u}'$ were Gaussian. There is no physical justification for this assumption and in fact it has led to disastrous results in practice. Such hypotheses overlook the realizability conditions imposed by lower order moments. Given n moments (i.e., averages of powers of the function), the $(n+1)^{st}$ must satisfy certain inequalities (the simplest example of such an inequality is $\overline{u} \leq \sqrt{\overline{u^2}}$). If the assumptions about the moments violate the inequalities, one is led to a problem with no solution. However, there is an infinite number of possible flows which can be obtained by closure approximations, and one would like to find a reasonable closure by considering the physics of turbulence.

Consider the ways in which laminar and turbulent flows differ. There are two major differences. The first is that in a turbulent flow there is a greatly heightened rate at which vorticity is produced at a boundary. In both laminar and turbulent flows, vorticity is produced at the boundary. In laminar flow, it diffuses into the fluid by molecular diffusion and by separation in a few well defined regions. In turbulent flow, the vorticity layer at the boundary is excessively thin, unstable, and is ejected into the fluid by processes which depend on its own dynamics; these processes are randomizing for poorly understood reasons.

The second major difference between laminar and turbulent flows is the greatly heightened dissipation of energy which occurs in a turbulent flow. Turbulence may even be viewed as that motion which a fluid must necessarily take to be able to dispose, through viscosity, of the increased amount of kinetic energy being given to it. That great agitation must arise is illustrated by the following argument: Consider the total kinetic energy of the fluid contained in D, i.e., $E = \frac{1}{2} \int \rho |\underline{u}|^2 \, dV$. If no external forces are acting on the fluid so as to increase its kinetic energy, then

$$\frac{dE}{dt} = -2\nu \int_D (\frac{\partial u_i}{\partial x_j} + \frac{\partial u_j}{\partial x_i})^2 \, dV \qquad (1)$$

which says, since $\nu > 0$, that the total kinetic energy
decreases (i.e., is converted into internal energy of molecular
motion) due to the work done in deforming fluid particles by the
viscous forces. As $\nu \to 0$, it becomes increasingly likely that
the flow is turbulent, yet also the rate of dissipation of energy,
$\left|\frac{dE}{dt}\right|$, increases. The only way in which a balance can be main-
tained in Equation (1) is if the gradients in the velocity field
increase, or the flow becomes more agitated.

In recent years it has become increasingly evident that
there is a great deal of structure to turbulent flows, and that
the presence and dynamics of this structure must be accounted
for in any turbulence theory. This order in the midst of chaotic
motion takes the form of regions of intense concentration of
vorticity; loosely speaking, vortex tubes and sheets. Boundary
layers may be considered to be highly unstable vortex sheets which
manifest their instabilities by producing vortex tubes, which lift
up from the surface by their own induced velocity field, and
become stretched by the rapidly moving outer flow. As vortex
tubes are stretched, their vorticity becomes confined to a
smaller region and is thus more concentrated, thereby inducing
a faster swirling motion of small physical dimensions. While
vorticity is only created at boundaries, once in a flow it has
the ability to collect together into tubes which then partici-
pate in the stretching process which drives energy to the small
scale motion.

The transfer of energy from the large to the small scale

motion through the process of vortex stretching is called an energy cascade. This is the process responsible for heightened dissipation since dissipation is greater at higher wave numbers. (See the form of the dissipation term $\nu \nabla^2 u$). This process is traditionally examined in wave number space, through the use of the energy density function, $E(k)$, defined through

$$E(k) = \int_{|k'| = k} \phi_{ii}(\underline{k})d\underline{k} \quad ,$$

where $\phi_{ij}(\underline{k})$ is the Fourier transform of the Eulerian velocity correlation function $R_{ij}(\underline{r}) = \overline{u_i'(\underline{x})u_j'(\underline{x+r})}$, and a repeated index implies summation. Note that $R_{ii}(0) = \Sigma\, \overline{u_i'^2}$ is the kinetic energy per unit mass. Then using the definition of $E(k)$ we have that

$$\text{total energy} = \frac{1}{2} \sum_i \overline{u_i'^2} = \frac{1}{2} \int_0^\infty E(k)dk \quad .$$

A plot of $E(k)$ gives a picture of the distribution of energy among the different scales of turbulent motion, i.e., the energy spectrum. A typical picture of $E(k)$ might be:

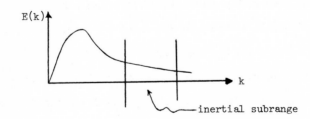

which shows that most of the energy is contained in the small
wave number region, i.e., is contained in the largest physical
eddies. One is to imagine the energy being put into the fluid
at small k , then cascading through smaller and smaller eddies
due to the mechanism of vortex stretching, until it is finally
dissipated at large wave numbers due to viscosity.

An important contribution to the theory of the cascade
process was made by Kolmogorov who suggested that at sufficiently
high Reynolds numbers there might exist an intermediate range of
k in which energy is not being dissipated or produced, but
only transferred to higher wave numbers. This range of k is
called the inertial subrange. If one postulates that the only
parameters determining E(k) are the total rate of energy
dissipation ε , and the wave number k , then through
dimensional analysis one must conclude that

$$E(k) \sim \varepsilon^{2/3} k^{-5/3} .$$

Kolmogorov had hoped that this result was universally true,
i.e., would hold for all turbulent flows. Unfortunately, it
does not appear today that this hypothesis is entirely correct.
Each turbulent flow that one encounters seems to have an energy
cascade which is partly its own.

The following argument, due to von Karman, suggests that
the turbulent flow near a boundary displays the features of the
Kolmogorov picture in physical space. Consider the flow:

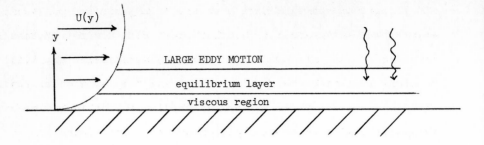

One may imagine that the energy produced in the outer flow, e.g.,
by the acceleration of the fluid due to pressure forces, cas-
cades through an equilibrium region (analogous to the inertial
subrange) to be dissipated in a small viscous region next to the
wall. von Karman hypothesized that the flow in this intermediate region would depend only on the distance from the wall, y ,
and on the shear stress, τ_W at the wall, and on the density
of the fluid ρ . Using dimensional analysis as before, one may
conclude that

$$\frac{dU}{dy} = K \frac{u^*}{y} \quad ,$$

where $u^* = \sqrt{\frac{\tau_W}{\rho}}$ is the friction velocity and K is a constant.
This relation may be solved for U to give the famous loga-
rithmic law of the wall:

$$\frac{U}{u^*} = K_1 \log \frac{yu^*}{\nu} + K_2 \quad , \quad K_1, K_2 \quad \text{constants}.$$

It is important to note what the Kolmogorov and von Karman arguments have in common: they both deal with vorticity; they are both vague; they are both unconcerned with the exact physical mechanisms which allow them to exist; and they both deal with the crucial range of scales in between the large visible scales and the small scales where dissipation occurs. This intermediate scale is the scale where chaotic behavior is expected, and thus one may hope that the ideas of dynamical system theory can shed light on the reasons for the partial validity of these two laws. The properties of these scales determine the "extra" unknowns $\overline{u_i' u_j'}$ in the averaged equations.

We now turn to the problem of incorporating such information about intermediate scales into a closure of the averaged Navier-Stokes equations. The simplest way to do this is through the mixing length or eddy viscosity models. By analogy with the kinetic theory of gases, one is led to assumptions of the form

$$- \overline{\rho u_1' u_2'} = \mu_e \frac{\partial \overline{u_1}}{\partial y}$$

or

$$- \overline{\rho u_1' u_2'} = \rho \ell^2 \left(\frac{\partial \overline{u_1}}{\partial y} \right)^2$$

where μ_e is an "eddy viscosity" and ℓ a "mixing length" (analogous in some ways to molecular viscosity and a mean free path). μ_e and ℓ are not constants, and are properties of the flow rather than of the fluid. One can exhibit constructions for μ_e and ℓ which ensure that the Kolmogorov law and the von Karman law are obeyed by the solutions of the resulting equations. Experiment is used to obtain additional information about reasonable choices for μ_e and ℓ .

There exist more sophisticated methods for closing the averaged Navier-Stokes equations. They all rely on experimental data, and have unclear physical and mathematical assumptions. What should ideally happen is that increased qualitative understanding of the dynamics of turbulence would lead to a more sophisticated use of experimental data in an increasingly plausible and useful closure system. The problems of vortex dynamics are crucial to this program; some progress has recently been made in this direction.

In summary, I showed that the averaged equations contain unknown terms which depend on small scale fluctations. I gave some of the most widely believed information about the properties of these small scales (those are the scales where dynamical system theory can be usually thought to be applicable), and I roughly outlined how this information can be used in making closures (i.e. finding equations with a number of unknowns small enough for the equations to be solvable). There is no generally accepted way of doing this last step, mostly because the problem is one of coupling scales with possibly different qualitative and mathematical properties.

REFERENCES

In addition to the standard references for this series of talks, readers may be interested in:

P. Bernard, Ph.D. Thesis, Berkeley, 1977.

P. Bradshaw, The understanding and prediction of turbulent flow, Aeronautical Journal, 1, 403 (1972).

A. J. Chorin, Numerical Study of Slightly Viscous Flow, J. Fluid Mech., 17, 785 (1973).

R. H. Kraichnan, The closure problem of turbulence theory, Proc. Symp. Applied Math., 13, 199 (1965).

H. Tennekes and J. L. Lumley, A First Course in Turbulence, M.I.T. Press (1972).

W. W. Willmarth, Structure of turbulent boundary layers, Adv. Applied Mech., 15, 159 (1975).

LECTURE III

DYNAMICAL SYSTEMS AND TURBULENCE

Steve Smale

The purpose of this talk is to present some questions and
ideas in the field of dynamical systems related to problems which
arise in turbulence. We shall begin with the discussion of how
the Navier-Stokes equations define a dynamics on a certain infinite
dimensional function space.

Recall that the law of motion of an incompressible viscous
fluid with constant density (this assumption has been made in both
previous talks) is given by the Navier-Stokes Equations:

$$\begin{cases} \frac{\partial v}{\partial t} - \nu \Delta v - (v \cdot \nabla)v = -\nabla p + f \\ \text{div } v = 0 \\ v = \text{prescribed on } \partial\Omega \end{cases}$$

where Ω is a region containing the fluid, v the velocity field
of the fluid, p the pressure and f the external forces. ν is
the kinematic viscosity, or, in the way we wrote the equations,
$\nu = 1/Re$, where Re is the Reynolds numbers. In all our talk,
Ω is supposed to be an open bounded set in $\mathbb{R}^3$ with smooth boundary;
$v:\bar{\Omega} \longrightarrow \mathbb{R}^3$ is also assumed to be "smooth". As was pointed out

already in lectures one and two, we believe that the chaoticness
in turbulence is intrinsically in the semiflow defined by the
Navier-Stokes equations.   But when dealing with a physical problem like this, one has to define a space of states.   We consider
S — the space of states of the physical system — to be the set of
all "smooth" maps $u : \bar{\Omega} \to \mathbb{R}^3$ with boundary conditions $u | \partial \Omega$ =
prescribed or tangent to $\partial \Omega$.   Consider now maps $u : \mathbb{R} \to S$,
where $\mathbb{R}$ is considered to be the time-axis and write $u_t(x) = u(t, x)$.
The determination of the map $u : \mathbb{R} \to S$, is given by the Navier-
Stokes Equations.   It is clear that we won't be satisfied with S
and will consider $S_0 = \{ u \in S \,|\, \mathrm{div}\ u = 0 \}$ and maps $u : \mathbb{R} \to S_0$.   In
this way, the Navier-Stokes equations define formally an ordinary
differential equation on S_0 .   For a discussion of the "smoothness"
of v as well as for candidates for S_0 (certain Sobolev spaces)
and how one copes with semiflows with unbounded generators, see
the Appendix to Lecture I.   Just let us stress here once again
the idea that the existence and uniqueness theorem for the solutions of the Navier-Stokes equations gives a dynamics on S_0
and one works with this dynamics.

There are at least two ways of attacking a discussion of the
semiflow defined by the Navier-Stokes Equations; each is based on
a finite dimensional approximation akin to the Galerkin Method.

I.   Find an invariant finite dimensional submanifold I, preferably
low dimensional.   The following might be a way to find such a submanifold: take the eigenvalue expansion of Δ and retain only
those eigenfunctions corresponding to strictly positive eigenvalues.

If there were no nonlinear term $(u \cdot \nabla)u$ one would obtain in this way an invariant linear manifold I attracting for the semiflow

We should obtain something perturbed, "around" I when considering the full Navier-Stokes Equations. This is a hope.

II. Take a finite dimensional space spanned by a finite number of eigenfunctions of the Laplacian and consider a projection

$$S_0 \longrightarrow \mathbb{R}^n .$$

On each $\mathbb{R}^n$ one gets via this projection a dynamical system induced from S_0. If $n \to \infty$ one hopes to be able to approximate the dynamics on S_0 with those defined earlier on $\mathbb{R}^n$.

Though mathematically challenging, both approaches miss something, namely, we should never forget that $I \subset S_0$ is in the space of vectorfields, so any result we obtain, cannot be physically tested under this form. The second question which arises is how can the two methods be tied together. How can a result in one of them mean something in the other? This should certainly happen, since both attempt to give information about the same physical phenomenon.

These two questions are answered by the introduction of observables, which are maps $g : S_0 \to \mathbb{R}$. For example one can consider

$g_x : S_0 \to \mathbb{R}^3$ defined by $g_x(u) = u(x)$ for each $x \in \Omega$. The main quality of these maps is, that their action can be actually physically tested, so one has a certain control over the mathematical results obtained earlier. Their second quality is that they tie the two approaches together. For the first one, we have a commutative diagram

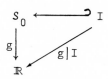

In the second approach one starts with maps h_n on the projections, $h_n : \mathbb{R}^n \to \mathbb{R}$ and obtains the observable h by the composition $S_0 \to \mathbb{R}^n \xrightarrow{h_n} \mathbb{R}$. The commutative diagram

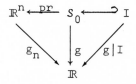

shows how an observable in one approach defines an observable in the second and vice-versa.

In the rest of my talk I shall be concerned only with questions of dynamical systems which inherently will appear when treating in the way described above the Naiver-Stokes Equations. I shall discuss two central questions related to this: stability and ergodicity.

Stability. Any reasonable mathematical model of a physical phenomenon should be "stable", i.e. if one makes certain perturbations, its qualitative features do not change. One must expect that, since any model represents an idealization of reality and hence represents itself a perturbation from the "real model", which of course is not present and nobody ever hopes to lay their hands on it. So, if our model couldn't be "stable", the derivation from reality can be disastrous and our model is no good!

When dealing with dynamical systems, the questions of stability become much more precise than the very vague principle stated above and they refer to the orbits. There are two concepts crucially related to stability in dynamical systems:

a) Attractor. We shall say that a set of orbits is an attractor or asymptotically stable if nearby orbits tend to the set as time increases; formally, an invariant closed set A of the flow F_t(i.e. $F_t A \subseteq A$) is an attractor if for any neighborhood U of A, there exists a neighborhood V of A such that if $x \in V$, then

$$\lim_{t \to \infty} \text{distance } (F_t(x),A) = 0.$$

b) Robust. We shall call those quantities robust which persist under slight perturbations of the system. Here we incorporate the various notions of stability found today in literature. One strong case of robustness is structural stability or invariance of the orbit structure under slight perturbations up to a continuous change of variables. Formally, two C^r-vectorfields $X,Y \in \mathfrak{X}^r$ (M) (M a compact manifold for example) are called topologically equivalent if there exists a homeomorphism $h:M \to M$ which sends the orbits of X onto the orbits of Y keeping their orientation, i.e. if $m \in M$ and $\varepsilon > 0$, there exists $\delta > 0$ such that for $0 < t < \varepsilon$,

$hF_t^X(m) = F_{t'}^Y(h(m))$ for some $0 < t' < \delta$, where F^X, F^Y denote the flows of X and Y respectively. $X \in \mathbf{X}^r(M)$ is said to be <u>structually stable</u> if there exists an open set 0 in the C^r-topology of $\mathbf{X}^r(M)$ such that all $Y \in 0$ are topologically equivalent to X.

Note that the concept of attractor refers to the invariance of the orbits relative to perturbations of the initial conditions whereas robutsness deals with insensitivity of the phase portrait under a perturbation of the system as a whole. Hence the "nicest" systems will be those which are robust in a region near an attractor not presenting qualitative changes at both types of perturbations.

<u>Example 1.</u> <u>Hyperbolic equilibrium</u> e.g.:

$$\begin{cases} \dfrac{dx}{dt} = x \\[2mm] \dfrac{dy}{dt} = -y \end{cases} \quad \text{with solutions} \quad \begin{cases} x = c_1 e^t \\[2mm] y = c_2 e^{-t} \end{cases}, \; c_1 \; c_2 \in \mathbb{R}$$

whose phase portrait looks like

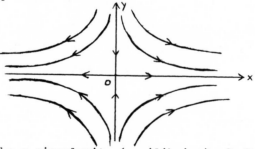

Let $m \in M$ be a singularity (equilibrium) of $X \in \mathbf{X}^r(M)$, i.e. $X(m) = 0$ where M is a compact manifold. We shall say that m is a <u>hyperbolic</u> <u>singularity</u> or <u>hyperbolic</u> <u>equilibrium</u> if $T_m X : T_m M \to T_m M$ has no eigenvalue with real part zero. Then it is known that the set of vectorfields which have all their singularities hyperbolic is open and dense in $\mathcal{X}^r(M)$ (see for example [PM], page 103). This shows that a hyperbolic equilibrium is robust since any nearby vectorfield has also only hyperbolic equilibria. Note that in this particular example robustness does not mean structural stability. Recall now the theorem which states that an equilibrium (i.e. singularity) is stable if and only if all eigenvalues of T_mX have strictly negative real parts; this is sometimes referred to as the "principle of linearized stability" Hence we would expect that there are hyperbolic unstable equilibria which indeed is the case as our above example shows.

Example 2. This will show that there are non-robust attractors. Consider in $\mathbb{R}^2$ the flow

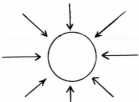

where the unit disc is the set of equilibria, all orbits outside the unit disc tend towards the unit circle. The unit disc is clearly an attractor, but is nonrobust, since a slight perturbation will destroy the bounded set of fixed points.

In classical differential equations one encounters two cases of compact attractors which are structually stable. The <u>first</u> <u>classical</u> <u>attractor</u> is an equilibrium $m \in M$ such that T_mX has all eigenvalues with strictly negative real part. The phase portraits look in this case like that, for example

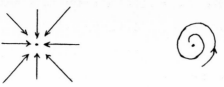

As concrete examples consider the equations in $\mathbb{R}^2$:

$$\begin{cases} \dfrac{dx}{dt} = x \\[2mm] \dfrac{dy}{dt} = y \end{cases} \quad \text{with solution} \quad \begin{cases} x = c_1 e^t \\[2mm] y = c_2 e^t \end{cases} \quad \text{for the first picture}$$

and

$$\begin{cases} \dfrac{dx}{dt} = -x - y \\[2mm] \dfrac{dy}{dt} = x - y \end{cases} \quad \text{with solution} \quad \begin{cases} x = e^{-t}(c_1 \cos t - c_2 \sin t) \\[2mm] y = e^{-t}(c_2 \cos t + c_1 \sin t) \end{cases}$$

for the second picture.

The <u>second classical attractor</u> is a stable attracting, closed orbit. For exaple, in Van der Pol's equation

$$\begin{cases} \dfrac{dx}{dt} = y - x^3 + x \\[2mm] \dfrac{dy}{dt} = -x \end{cases}$$

one finds a unique closed orbit, all solutions from outside

spiral towards it and all solutions from inside expand towards it, the origin being a source (for a study of Van der Pol's equation, see [HS], p. 215-228).

We shall now describe a new kind of nonclassical structurally stable attractor not found in the traditional theory of ordinary differential equations and I shall call such attractors <u>strange</u>

attractors. Since everything will be done by means of discrete dynamical systems, I mention at this point that there is a canonical way to associate to each discrete dynamical system on a (compact) manifold M a global flow — hence a vectorfield — on a manifold M_0 of one dimension higher. This then will show that we did not restrict the generality of the example by working — easier — with discrete dynamical systems. The general theory will be presented at the end. Our exposition follows Shub's paper [Sh].

We start off with the "expanding" map $f:S^1 \to S^1$ given by $f(z) = z^2$. If D denotes the full disc in 2-dimensions, denote by $R = S^1 \times D$ the full ring having as boundary the two dimensional torus. For matters of convenience we shall imagine the ring R centered at the origin of $\mathbb{R}^3$ with S^1 embedded in R as the central circle lying in the xOy plane and having radius equal to 2. Put

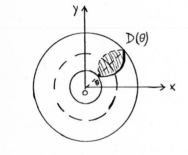

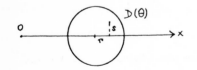

now coordinates (θ,r,s) in this ring in the following way: θ measures the angle in trigonometric sense from the Ox axis; (θ,r) forms a coordinate system on the annular region $R \cap (xOy)$ such that on the central circle of radius 2, $r = 0$; (r,s) forms a coordinate system in each "slice" of the ring, s measuring the "height" of the point in the particular "slice" $D(\theta)$ so that we always have $r^2 + s^2 \leq 1$. In this coordinate system, our map f looks like $(\theta,0,0) \to (2\theta,0,0)$. We want now to find an embedding $h:R \to R$ which models this "twice wrapping" of the central circle. Define $h:R \to R$ by

$h(\theta,r,s) = (2\theta, \varepsilon_1\cos\theta + \varepsilon_2 r, \varepsilon_1\sin\theta + \varepsilon_2 s)$ for $\varepsilon_2 < \varepsilon_1 < 1/2$; it is easy to check that h is in fact an embedding of R into itself. Our "slice" $D(\theta)$ is a cross-section of the ring through the point $(\theta,0,0)$ on the central circle, notice now that $h(D(\theta/2))$, $h(D(\theta/2+\pi)) \subset D(\theta)$. Actually $h(D(\theta(2)))$, $h(D(\theta/2+\pi))$ are small discs. We show that now and also compute their centers and radii. Their centers are $h(\theta/2,0,0) = (\theta, \varepsilon_1\cos\theta/2, \varepsilon_2\sin\theta/2)$, $h(\theta/2 + \pi,0,0) = (\theta, -\varepsilon_1\cos\theta/2, -\varepsilon_1\sin\theta/2)$ so that the distance between them — in the $D(\theta)$ plane of course — is $2\varepsilon_1$. To find the radius take an arbitrary point (θ,r,s) on the boundary, i.e. $r^2 + s^2 = 1$ and observe that $h(\theta/2,r,s) = (\theta, \varepsilon_1\cos\theta/2 + \varepsilon_2 r, \varepsilon_1\sin\theta/2 + \varepsilon_2 s)$ is at distance ε_2 from the center $(\theta, \varepsilon_1\cos\theta/2, \varepsilon_1\sin\theta/2)$. Similarly the radius of the other disc is ε_2. It is clear that our

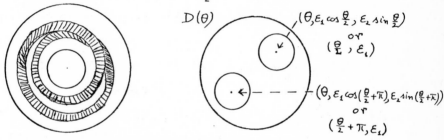

map h wraps our initial ring R by making it thinner and longer twice around the central hole of R, inside R. The phenomenon that occured was: in the θ plane two new discs of radius ε_2 appear with the centers at $(\theta, \varepsilon_1\cos\theta/2, \varepsilon_1\sin\theta/2)$ and at $(\theta, \varepsilon_1\cos(\theta/2 + \pi), \varepsilon_1\sin(\theta/2 + \pi))$. Write now formally (φ,ε) for $(\varphi, \varepsilon\cos\varphi, \varepsilon\sin\varphi)$ so that with this convention our centers will be at $(\theta/2,\varepsilon_1)$ and $(\theta/2 + \pi,\varepsilon_1)$ respectively. The second iterate has the analytic expression

$$h^2(\theta,r,s) = (4\theta, \varepsilon_1\cos 2 + \varepsilon_1\varepsilon_2\cos\theta + \varepsilon_2^2 r \ , \ \varepsilon_1\sin 2\theta + \varepsilon_1\varepsilon_2\sin\theta + \varepsilon_2^2 s).$$

As before, a computation shows that $h^2(D(\theta/4))$, $h^2(D(\theta/4 + \pi)) \subset$

$\subset h(D(\theta/2)) \subset D(\theta)$, $h^2(D(\theta/4 + \pi/2))$, $h^2(D(\theta/4+3\pi/2)) \subset$

$\subset h(D(\theta/2+\pi)) \subset D(\theta)$; our picture then looks like:

The four small discs which appear have radius ε_2^2 and the centers —
using our symbolical writing — at:

$$(\tfrac{\theta}{2},\varepsilon_1) + \varepsilon_2(\tfrac{\theta}{4},\varepsilon_1); \ (\tfrac{\theta}{2},\varepsilon_1) + \varepsilon_2(\tfrac{\theta}{4}+\pi,\varepsilon_1)$$

$$(\tfrac{\theta}{2}+\pi,\varepsilon_1) + \varepsilon_2(\tfrac{\theta}{4}+\tfrac{\pi}{2},\varepsilon_1); \ (\tfrac{\theta}{2}+\pi,\varepsilon_1) + \varepsilon_2(\tfrac{\theta}{4}+\tfrac{3\pi}{2},\varepsilon_1) \ .$$

This procedure then goes indefinitely: at each stage, for each
earlier obtained disc, 2 new discs with radius shrunk by a factor
of ε_2 will appear inside the older discs. That gives us the idea
of looking at the whole process differently: at the first stage
write pairs $(\theta_1,\theta_2) \in S^1 \times S^1$ such that $2\theta_2 = \theta_1$; that corresponds
to the idea that the first coordinate gives the θ for the slice
$D(\theta)$ and the second coordinate tells us for which angle, after
applying h we still land in the chosen "slice". For the second
iterate we write triples $(\theta_1,\theta_2,\theta_3) \in S^1 \times S^1 \times S^1$ with $2\theta_2 = \theta_1$,
$2\theta_3 = \theta_2$; the first coordinate gives the angle of the chosen slice,
the second gives the angle for which the first iteration yields the
2 small discs in the chosen slice and the third coordinate shows
for which angle the second iteration lands in the smaller discs.
Note that the knowledge of this sequence plus the fact that at each

stage we shrink ε_2 times the surface of the section and the pattern of formation of the centers of the consecutive discs tells us everything we need to know in order to locate any disc obtained in this process. Generally, we take the infinite product $S^1 \times \ldots \times S^1 \times \ldots$ and form the subset $\{(\theta_1, \ldots, \theta_k, \ldots) \mid 2\theta_{k+1} = \theta_k$ for all $k \geq 1\}$. It is clear that each such sequence determines a unique point of $\bigcap_{n \geq 0} h^n(R)$ and conversely. Our space of sequences will be called <u>solenoid</u> and what we showed is that our solenoid is homeomorphic to $\bigcap_{n \geq 0} h^n(R)$ (the verification of continuity is easy and will be left to the reader). Note that $\bigcap_{n \geq 0} h^n(R)$ is in each disc $D(\theta)$ a Cantor set. Also, we remark that $\Lambda = \bigcap_{n \geq 0} h^n(R)$ attracts all points of R under the iterates of h and that it is locally a product of a Cantor set and a one-dimensional arc. More, at each point of Λ we have a so called "hyperbolic splitting" which has to be understood by the fact that h is expanding in one direction (makes R always longer) and contracting in other two dimensions (always shrinks the section by a factor of ε_2). It is proved that Λ is robust, in this particular case structurally stable: small C^1-perturbations of the map preserve this picture. Exact mathematical statements in a general framework will be provided at the end of this lecture.

Let us return now to our solenoid and present R. Williams' idea of studying such a strange attractor by inverse limits; much more about this will be said in his later talk. Recall that our solenoid was the set $\{(\theta_1, \ldots \theta_k, \ldots) \in S^1 \times \ldots \times S^1 \times \ldots \mid \theta_k = 2\theta_{k+1}\}$ which is obviously a closed subspace of $S^1 \times \ldots \times S^1 \times \ldots$ endowed with the product topology. Our solenoid is actually an inverse limit of a certain inverse system. To see this, denote $X_i = S^1$ and define $u_{ij}:X_j \to X_i$ for $i < j$, $i,j \in \mathbb{N}$ by $u_{ij}(\theta_j) = 2^{i-j}\theta_j$; in particular $u_{k,k+1}(\theta_{k+1}) = \theta_{k+1}/2$. Then

$$\varprojlim_{\mathbb{N}}(X_i,u_{ij}) = \{(\theta_1,\ldots,\theta_k,\ldots) \in X_1 \times \ldots \times X_k \times \ldots \mid u_{ij}(\theta_j) = \theta_i = 2^{i-j}\theta_j \, \forall_i < j\} =$$

$$= \{(\theta_1,\ldots,\theta_k,\ldots) \in S^1 \times \ldots \times S^1 \times \ldots \mid 2\theta_{k+1} = \theta_k, \, \forall k \in \mathbb{N}\}, \text{ which is}$$

our solenoid. $f|\Lambda$ via our identification becomes just the shift $(\theta_1,\ldots,\theta_k,\ldots) \longmapsto (2\theta_1,\theta_1,\ldots,\theta_k,\ldots)$. Even though our construction of the inverse limit here might seem a somewhat forced formalism, this idea is extremely fruitful when dealing with semiflows on branched manifolds (see R. Williams' talk).

Ergodicity. Accomplishing our first goal by giving an example of a strange attractor, we shall concentrate on the second important feature of attractors called ergodicity. Loosely speaking ergodic theory is the study of transformations and flows from the view point of measure theory. We leave the exact definitions and theorems for the end of the talk and pursue our previous example in order to get some feeling of what ergodic theory migh provide for the study of attractors of a diffeomorphism.

We start with the remark that if $x,y \in D(\theta)$, then $d(h^n(x),h^n(y)) \to 0$ so that if $g:R \to \mathbb{R}$ is a continuous map we shall have $|gh^n(x) - gh^n(y)| \to 0$ and hence the time averages approach each other:

$$\frac{1}{n}\left|\sum_{i=0}^{n-1}gh^i(x) - \sum_{i=0}^{n-1}gh^i(y)\right| \to 0 .$$

This has to be interpreted in the following way: one limit exists

if and only if the other does. It is known in our case that these
limits exist and satisfy

$$\lim_{n \to \infty} \frac{1}{n} \sum_{i=0}^{n} gh^i(\theta,r,s) = \int g(\theta,r,s)d\theta$$

for almost all (θ,r,s) in R. Technically this says that the usual
Lebesgue measure on S^1 is ergodic and invariant with respect to our
transformation h and our statement is a typical ergodic theorem.

We define now a measure μ on our strange attractor Λ in the
following way: for each continuous $g:\Lambda \to \mathbb{R}$ put $\hat{g}(\theta) = \min \{g(\theta,r,s)|$
$(\theta,r,s) \in D(\theta)\}$ and by the Riesz representation theorem we can
find a unique measure μ on Λ such that $\int_{\Lambda} gd\mu = \int_{S^1} \hat{g}d\theta$. Then it is
proven that μ is an invariant measure which is ergodic with re-
spect to h and for almost all $(\theta,r,s) \in R$ we have
$\lim_{n \to \infty} \frac{1}{n} \sum_{i=0}^{n-1} gh^i(\theta,r,s) = \int_{\Lambda} gd\mu$. We end these short comments about
ergodic theorems with the remark that this last statement is strongly
related to our above inverse limit construction and to the fact that
via this construction we transformed h on the attractor to a shift
in a certain sequence space. These ideas have been beautifully developed by Bowen and the results he gets — for example in [Bo] and
[BoR] — using these techniques contributed very much to the understanding of ergodic properties of certain types of attractors.

The most popular invariant in ergodic theory is <u>entropy</u>. We
shall abstain here from exact definitions and only mention that the
topological entropy (that's the one we are talking about here) essentially gives the asymptotic exponential growth rate of the number
of orbits of a certain diffeomorphism up to any accuracy and arbitrary

high period. Very loosely speaking and disregarding an unproved conjecture, the entropy of a diffeomorphism is bounded below by the logarithm of how many times it "wraps around" the manifold. In our case, the entropy will be log 2, that is log deg h and this is by no means an accident.

We refer the reader to R. Bowen's talk for much more detailed information about ergodicity.

General Theory. This section aims to present roughly the general mathematical machinery behind our previous construction of a strange attractor. We shall begin with an earlier promised result namely the explicit construction of a flow on a manifold of one dimension higher from a discrete dynamical system, it is done after [S] page 797.

Given a vectorfield X on the manifold M, a cross-section of X is a closed codimension one submanifold Σ of M such that every orbit of X intersects Σ , Σ is transverse to the flow of X and every orbit leaving Σ intersects Σ in both future and past time If this happens, define the first return map $f:\Sigma \to \Sigma$ by setting $f(x) = F_{t_0}(x)$, where F_t is the flow of X and $t_0 > 0$ is the least t_0 with $F_{t_0}(x) \in \Sigma$. By the global smoothness of the flow, f is a smooth map. Note that the orbits of X are in this way in one to one correspondence with the orbits of f, i.e. with $\{f^k(x) | k \in$ compact orbits are preserved under this correspondence and hence periodic points of f correspond to closed orbits of X. However, the existence of a cross-section is not always guaranteed; for example, X cannot have singularities and this will then restrict by the Poincaré-Hopf index theorem for compact manifolds the topological type of M.

On the other hand, the converse construction is much nicer, namely, every diffeomorphism can be regarded as the first return map of a cross section of some flow. Given the discrete dynamical system defined by the diffeomorphism $f:M \to M$ define the map $\alpha: \mathbb{R} \times M \to \mathbb{R} \times M$ by $\alpha(s,m) = (s+1, f(m))$. It is clear that $\mathbb{Z}$ operates freely on $\mathbb{R} \times M$ via the action $(k,(t,m)) \to \alpha^k(t,m) = (s+k, f^k(m))$ so that the orbit space $(\mathbb{R} \times M)/\mathbb{Z} = M_0$ is a manifold of dimension $1 + \dim M$. Note that the flow $\phi_t: \mathbb{R} \times M \to \mathbb{R} \times M$ defined by $\phi_t(s,m) = (s+t,m)$ is $\mathbb{Z}$-equivariant i.e. $\phi_t \circ \alpha^k = \alpha^k \circ \phi_t$ for all $k \in \mathbb{Z}$, $t \in \mathbb{R}$, so that it induces a flow $F_t:M_0 \to M_0$ defined by $F_t(\mathbb{Z}(s,m)) = \mathbb{Z}(\phi_t(s,m)) = \mathbb{Z}(s+t,m)$ where $\mathbb{Z}(s,m)$ denotes the $\mathbb{Z}$-orbit through $(s,m) \in \mathbb{R} \times M$. By a general principle of passing to quotients, F_t is smooth (see [B], page 51) and we have defined in this way a smooth global flow $F:\mathbb{R} \times M_0 \to M_0$ called the <u>suspension</u> of f. We shall emphasize some remarkable properties of the suspension. Note that if $\pi:\mathbb{R} \times M \to M_0$ is the canonical projection — which is a submersion — then $\Sigma = \pi(0 \times M)$ is a cross-section of the flow F_t diffeomorphic via π to M. Also, for each $\mathbb{Z}(0,m) \in \pi(0 \times M)$, $F_1(\mathbb{Z}(0,m)) = \mathbb{Z}(1,m) = \mathbb{Z}(0,f^{-1}(m)) \in \Sigma$ and $t = 1$ is the smallest $t > 0$ for which this happens. Because of the commutative diagram

$$
\begin{array}{ccc}
M \ni m & \longmapsto & f(m) \in M \\
\Big\downarrow{\scriptstyle \pi} & & \Big\downarrow{\scriptstyle \pi} \\
\Sigma \ni \mathbb{Z}(0,m) & \xrightarrow{\ F_1\ } & \mathbb{Z}(1,m) \in \Sigma
\end{array}
$$

with vertical arrows diffeomorphisms, our initial f is conjugated
to the first return map of the new constructed flow. This construction has a certain property which makes it canonical. If $G:\mathbb{R} \times M \to M$
has a cross section $g: \Sigma \to \Sigma$ whose suspension is $F:\mathbb{R} \times M_0 \to M_0$
then G and F are equivalent by an orbit preserving homeomorphisms.

After this lengthy digression we return to the construction of
a strange attractor for a diffeomorphism on a compact manifold M.
It should be pointed out that Ruelle and Takens suggested that these
attractors might lead to turbulence, namely, they say that from an
equilibrium via a Hopf bifurcation a closed orbit is formed and the
successive higher order bifurcations (or other mechanism, as in the
Lorenz equations (Lecture I)) eventually lead to strange attractors
reference is made to the example described before (see [RT], page
170-171).

First we need some definitions and statement of theorems basic
in dynamical systems. If E is a Riemannian vector bundle over M
and $f:E \to E$ a vector bundle morphism, f will be called <u>contracting</u>
(<u>expanding</u>) if there exists $c > 0$, $0 < \lambda < 1$ ($\lambda > 1$) such that for
all $v \in E$, $n \in \mathbb{N}$, $\|f^n(v)\| > c\lambda^n \|v\|$ (respectively $\|f^n v\| > c\lambda^n \|v\|$).
Our definition obviously depends on the norms induced by the
metrics; however, these norms are equivalent locally and hence if we
suppose in addition that M is compact, the property of f being
contracting or expanding is independent of the Riemannian metric on
M. It is also clear that the inverse of an expanding bundle auto-
morphism is contracting and vice-versa. Now, if M is a compact
Riemannian manifold $f:M \to M$ will be called <u>contracting</u> (<u>expanding</u>)
if $Tf: TM \to TM$ is contracting (expanding).

From now on $f: M \to M$ will always denote a diffeomorphism on a compact manifold M.

A closed subset Λ of a compact manifold M is <u>hyperbolic</u> if $f(\Lambda) = \Lambda$, the tangent bundle of M restricted to Λ, $T_\Lambda M$ is a continuous Whitney sum $T_\Lambda M = E^s \oplus E^u$ so that for each $x \in \Lambda$, $Tf(E_x^s) = E_{f(x)}^s$, $Tf(E_x^u) = E_{f(x)}^u$ and Tf is contracting on E^s and expanding on E^u, i.e. there exist constants $c > 0$, $\lambda \in (0,1)$ such that $\|Tf^n(v)\| \leq c\lambda^n \|v\|$ for all $v \in E^s$, $n \geq 0$, $\|Tf^{-n}(v)\| \leq c\lambda^n \|v\|$ for all $v \in E^u$, $n \geq 0$. Note that in this definition the constants c and λ do depend on the used Riemannian metric even though the property of Λ being hyperbolic is independent of the metric. We shall come back to these constants a little bit later.

A point $x \in M$ is called wandering if there is a neighborhood U of x in M such that $f^n(U) \cap U = \phi$ for all $n > 0$. The set of <u>wandering points</u> is clearly open and invariant under f so that $\Omega(f)$, <u>the set of nonwandering points</u> is closed and f-invariant. $x \in M$ is <u>periodic</u> if $f^n(x) = x$ for some $n > 0$; clearly such an x is in $\Omega(f)$.

<u>Axiom A.</u> $f: M \to M$ satisfies Axiom A if $\Omega(f)$ is hyperbolic and $\Omega(f) = \{x \in M \mid x \text{ is periodic}\}$.

A metric is <u>adapted</u> to an Axiom A diffeomorphism f, if $\Omega(f)$ is hyperbolic with respect to it with $c = 1$. It is proved in [HP] that every Axiom A diffeomorphism has an adapted metric so that from now on whenever we deal with Axiom A diffeomorphisms we suppose to have already an adapted metric on M.

<u>Spectral Decomposition Theorem.</u> If $f: M \to M$ is an Axiom A diffeomorphism on the compact manifold M, then there is a unique way of

writing $\Omega(f) = \Omega_1 \cup ... \cup \Omega_k$ where the Ω_i are pairwise disjoint closed sets each one of them containing a dense orbit of f (i.e. f is topologically transitive) and such that $f(\Omega_i) = \Omega_i$ (i.e. Ω_i are f-invariant). More, $\Omega_i = X_{1,i} \cup ... \cup X_{n_i,i}$ with the $X_{j,i}$'s pairwise disjoint closed sets, $f(X_{j,i}) = X_{j+1,i}(X_{n_i+1,i} = X_{1,i})$ and $f^{n_i}|X_{j,i}$ topologically mixing.

Recall that a homeomorphism $h:N \to N$ is <u>topologically mixing</u>, if for any two open sets U,V of N, there exists $n_0 \geq 0$ such that for all $n \geq n_0$, $U \cap h^n(V) \neq \phi$. The proof of this strong version of the spectral decomposition theorem can be found in [Bo], page 72-74. Denote $W^s(\Omega_i) = \{x \in M | f^n(x) \to \Omega_i$ as $n \to \infty\}$ and note that Ω_i is an attractor if and only if $W^s(\Omega_i)$ contains a neighborhood of Ω_i. Ω_i are called <u>basic sets</u>.

Theorem (Bowen, Ruelle). Let m denote the measure defined by the Riemannian metric on M. Then a basic set Ω_i of a C^2 Axiom A diffeomorphism is an attractor if and only if $m(W^s(\Omega_i)) > 0$. (For the proof see [BoR], page 195).

This result coupled with the spectral decomposition theorem yields the following

Corollary. Every C^2 Axiom A diffeomorphism has at least an attractor. Almost every point of M tends to an attractor under iterates of f.

Also note that for Axiom A diffeomorphisms we have $M = \bigcup_{i=1}^{k} W^s(\Omega_i$

We shall return to the basic sets investigating measure theoretical properties when dealing with ergodic properties later on.

With these results in mind, we can describe the general principle of construction of strange attractors as it is presented in [S], page 788. We start off with an expanding diffeomorphism $f:M \to M$ of a compact manifold. Denote by D the full unit disc of dimension $1 + \dim M$ and imbed M in $D \times M$ as $0 \times M$. For $0 < \lambda < 1$ define $g_\lambda : D \times M \to D \times M$ by $g_\lambda(x,y) = (\lambda x, y)$. Look now at the map $0 \times M \in (0,y) \to (0,f(y)) \in D \times M$; since $\dim(D \times M) = 1 + 2\dim M$ and M is compact, the imbeddings form a dense subset of $C^1(0 \times M, D \times M)$ in the C^1-strong topology (see [H], page 55) so that our map has a C^1-approximation $\varphi : 0 \times M \to D \times M$ which is an imbedding. Let now T be a tubular neighborhood of $\varphi(M)$ with fibers being the various components of $T \cap (D \times \{y\})$ for $y \in M$. Now extend φ to $\psi : T \to D \times M$ such that ψ is a diffeomorphism and is fiber preserving. Pick now λ small enough such that $g_\lambda(D \times M) \subseteq T$; this can be done by a compactness argument. Now define for these λ's $h = \psi \circ g_\lambda : D \times M \to D \times M$. The following facts hold: $\Lambda = \bigcap_{m > 0} h^m(D \times M)$ has a hyperbolic structure, is a basic set for h and is structurally stable; Λ is locally the product of a Cantor set and a manifold of dimension equal to $\dim M$.

Williams' construction in this general setting repeats step by step what we've done in the particular case $M = S^1$. Namely, denoting $X_i = M$, define $u_{ij} : X_j \to X_i$ for $i < j$ by $u_{ij}(m_j) = f^{i-j}(m_j)$ and notice that

$$\varprojlim_{\mathbb{N}}(X_i, u_{ij}) = \{(m_1, \ldots, m_k, \ldots) \in X_1 \times \ldots \times X_n \times \ldots \mid u_{ij}(m_j) = m_i =$$
$$= f^{i-j}(m_j), \forall i < j\} = \{(m_1, \ldots, m_x, \ldots) \in M \times \ldots \times M \times \ldots \mid f(m_{k+1}) = m_k\},$$

the last definition of this space being extremely suggestive for our process of forming Λ. Actually, the obvious map given by the construction of Λ , $\Lambda \ni (x,m) \mapsto (m,f^{-1}(m),\ldots,f^{-k}(m),\ldots, \in \varprojlim_{\mathbb{N}}(X_i,u_i$

establishes an isomorphism of Λ with our above defined solenoid. Via this identification, h becomes the shift:

$(m_1,\ldots,m_k,\ldots) \mapsto (f(m_1),m_1,\ldots,m_k,\ldots)$.

Strongly related to this construction is the following:

Theorem (Bowen, Ruelle). Let f be a C^2 Axiom A diffeomorphism on M and Ω_i a basic set. Then for m - almost all points $x \in W^s(\Omega_i)$ one has

$$\lim_{n \to \infty} \frac{1}{n} \sum_{i=0}^{n-1} gh^i(x) = \int_{\Omega_i} gd\mu$$

for all continuous $g:M \longrightarrow \mathbb{R}$. Here μ denotes a certain probability measure on Ω_i, invariant under h which has remarkable ergodic properties.

For the construction of μ , a proof and other related results see [BoR], page 191.

Topological entropy is defined by Bowen in the following way. Let $f:M \to M$ be a continuous map on a compact metric space M. For given $\varepsilon > 0$, $n \in \mathbb{N}$, a set $E \subset M$ is called (ε,n)-separated if for any $x,y \in E$ with $x \neq y$ there is a j, $0 \leq j \leq n$ such that $d(f^j(x),f^j(y)) > \varepsilon$. It is easily seen that such a set E must be discrete and closed, hence finite by compactness of M. Denote by $Z_n(f,\varepsilon)$ the largest cardinality of any (n,ε)-separated set in M and let $h(f,\varepsilon) = \lim \sup_{n \to \infty} \frac{1}{n} \log Z_n(f,\varepsilon)$. The topological entropy of f is by definition: $h(f) = \lim\limits_{\varepsilon \to 0} h(f,\varepsilon)$. Denote by $N_m(f)$ the cardinality of the fixed point set of f^m, i.e. $N_m(f)$ is the number of periodic points of f of period m. Then Bowen proves (see [Bol] for a proof)

$$h(f) = \limsup\limits_{m \to \infty} \frac{1}{m} N_m(f)$$

whenever f is an Axiom A diffeomorphism on the compact manifold M.

This theorem then justifies the entropy statement in our example. Indeed, in order to have $h^n(\theta,r,s) = (\theta,r,s)$ we must map $D(\theta)$ into itself by h^n and once we establish the conditions on θ, the coordinates r and s follow automatically from our condition. So we really have to ask how many periodic points of period n does $S^1 \ni z \mapsto z^2 \in S^1$ have; the answer is clearly $2^n - 1$. Now remark that our diffeomorphism h satisfies Axiom A ($\Omega(f) = \Lambda$) so that its topological entropy equals $\lim\limits_{n \to \infty} \frac{2^n - 1}{n} = \log 2$. We didn't define here the usual measure theoretical entropy which differs in general from the topological entropy; however the two concepts are strongly related. We refer the reader to Bowen's talk for more information on ergodicity.

I finish this talk with the remark that the Lorenz attractor described in Lecture I doesn't fit into this general framework of strange attractors. Attractors like these haven't been yet systematized.

BIBLIOGRAPHY

[B]   Bourbaki, N.   Variétés Différentiables, Fascicule des
       Résultats, §1-8, Hermann, Paris

[Bo]   Bowen, R.   Equilibrium States and the Ergodic Theory of
       Anosov Diffeomorphisms, Lecture Notes in Mathematics
       470, Springer Verlag 1975.

[Bo1] Bowen, R.   Topological Entropy and Axiom A, Proc. Symp.
       Pure Math., vol. 14, Amer. Math. Soc. Providence R.I.
       1970, pp. 23-41.

[BoR] Bowen, R., Ruelle, D.   The Ergodic Theory of Axiom A Flows,
       Inventiones Math., 29 (1975) pp. 181-202.

[H]   Hirsch, M.   Differential Topology, Graduate Texts in
       Mathematics 33, Springer Verlag 1975.

[HS]   Hirsch, M., Smale, S.   Differential Equations, Dynami-
       cal Systems, and Linear Algebra, Academic Press, 1974.

[HP]   Hirsch, M., Pugh, C.   Stable Manifolds and Hyperbolic
       Sets, Proc. Symp. Pure Math. 14 (1970), 133-163.

[PM]   Palis, J., deMelo, W.   Introdução aos Sistemas Dinâmicos,
       Colóquio Brasileiro de Matematica Poços de Coldas,
       Julho 1975, IMPA.

[R]   Robbin, J.   Topological Conjugacy and Structural
       Stability for Discrete Dynamical Systems, BAMS, vol. 78,
       No. 6, November 1972, pp. 923-952.

[RT]   Ruelle, D., Takens, F.   On the Nature of Turbulence,
       Commun. Math. Phys. 20 (1971), pp. 167-192.

[S]   Smale, S.   Differentiable Synamical Systems, B.A.M.S.
       73 (1967) pp. 797-817.

[Sh]   Shub, M.   Stability in Dynamical Systems, (preprint).

[Sh1] Shub, M.   Dynamical Systems, Filtrations and Entropy,
       B.A.M.S.   vol. 80, No. 1, January 1974, pp. 27-41.

[W]   Walters, P.   Ergodic Theory — Introductory Lectures,
       Lecture Notes in Mathematics 458, Springer Verlag,
       1975.

[Wi]   Williams, R.   Expanding Attractors, Publications
       Mathematiques, no. 43, I HES, 1974.

LECTURE IV

A PHENOMENONOLOGICAL THEORY FOR THE COMPUTATION OF
TURBULENT SHEAR FLOWS

P. G. Saffman

The subject of turbulence may be considered to be the study
of random solutions of the Navier-Stokes equations, subject to
specified initial and boundary conditions. Precisely what is meant
by 'random' depends on one's point of view. To a mathematician,
"randomness" implies the existence of an ensemble of possible
realizations of the flow field on which averages are defined.
To an engineer, each turbulent flow situation that is encountered
in practice is actually one single continuously occuring reali-
zation of the flow field and "random" really means "unpredictable".
The engineer hopes to discover predictable properties of the flow
field by suitable time and/or spatial averaging of the flow field.

Historically, two different types of turbulence have been
investigated: imaginary turbulence and real turbulence. Imaginary
turbulence is an idealized turbulent motion, not found in nature,
which is both homogeneous and isotropic. "Homogeneous" means that
the mean flow properties are spatially uniform, while the condition of isotropy means that there is no preferred orientation to

the averaged properties of the turbulent fluctuations. The turbulent flows which do occur in nature may be termed 'real', and they are invariably non-homogeneous.

Turbulence which is approximately homogeneous and isotropic does exist, and in fact can be generated in a laboratory wind tunnel by passing a fluid stream of sufficiently high velocity past a fine mesh. The turbulent fluid motion far enough downstream of the grid, is found experimentally to be roughly isotropic, and if one moves with the mean flow velocity, the turbulence will also appear to be homogeneous. The averaged properties of homogeneous and isotropic turbulence may vary with time, and in fact, the rate of decay of turbulent kinetic energy of such flows has been the object of considerable study.

The theoretical investigation of homogeneous and isotropic turbulence was begun by Taylor, and it was originally throught that this kind of turbulent motion might be a less formidable problem to study mathematically than real turbulence. Unfortunately, the mathematical difficulties to be faced in imaginary turbulence are still very great, and the amount of understanding that has been obtained about such motion is limited. It is also not clear that there is any benefit to be gained from the study of this kind of turbulence, since the presence of shear in all real turbulence flows has a non-trivial effect on the dynamics of the turbulent motion at all scales.

Real turbulent flows may be loosely divided into two types: those which are 'simple' and those which are 'complex'. Simple

turbulent flows are those which are devoid of extraneous geometrical or phenomenonological complications, while complex turbulent
flows are those which are not simple. For example, compare the
turbulent flow in a pipe:

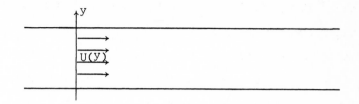

Figure 1

in a jet:

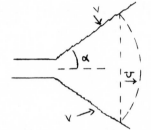

Figure 2

or in a mixing layer:

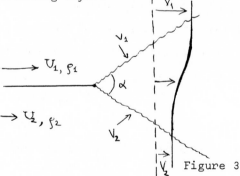

Figure 3

all of which we will call simple turbulent flows, with the turbulent

flow past a cylinder:

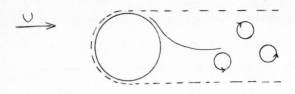

Figure 4

or in a turbulent mixing layer near a corner:

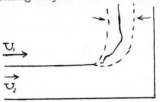

Figure 5

both of which are examples of complex turbulent flows.

All of these examples, both simple and complex, have defied

understanding by theoretical means. Consider the turbulent flow

in a pipe, shown in Figure 1. For the region $\frac{\nu}{U*} \ll y \ll a$, where U

is the kinematic viscosity, $u* = \sqrt{\tau_w/\rho}$ is the friction velocity,

τ_w is the shear stress at the wall, and a is proportional to

the pipe diameter, the logarithmic law of the wall for the mean

velocity distribution:

$$U(y) = U*\left[\frac{1}{K} \log \frac{U*y}{\nu} + B\right]$$

has been found to hold for turbulent flow in any pipe at sufficiently high Reynolds numbers. K = 4.2 is the Karman constant,
which is the same for all pipes, while B is a constant which
varies depending on the roughness of the pipe walls. This relation, which in fact holds for the turbulent flow past any surface,
has never been derived by rigorous theoretical arguments.

For the turbulent flow due to a jet of fluid leaving an
orifice, shown in Figure 2, one would like to be able to predict
apriori, the angle of spread of the jet, α , which for many different flow situations has been observed to be ~ 15° . Also the
rate of entrainment of the outer fluid into the jet is an important quantity to predict and no theory of turbulence is able
to do this.

Similarly, for the mixing layer in Figure 3 one would again
like to predict the spread angle and the entrainment velocities
V_1 and V_2 shown there.

For the turbulent wake behind a cylinder shown in Figure 4,
one would like to predict the characteristics of the turbulent
boundary layer adjacent to the cylinder, and the separation region
immediately behind it. Also one would hope that a theory of
turbulence could predict the presence of a row of large eddies which
are shed alternately from the flow above and below the cylinder.

Variations of the complex flow situation shown in Figure 5 are
frequently encountered in practice, and one would like to predict
how the mixing layer shown there is affected by the presence of
the corner, i.e. how it is bent, and the rate of entrainment of
fluid from the external flow into it.

We must caution that the simplicity of the flows shown in the first three figures is more apparent than actual. For example, instantaneous photographs of the flow in a jet by Roshko, reveal the intricate regular structure, shown in Figure 6.

Figure 6

The eddies depicted there are in a continuous process of combining and distorting, in a remarkably complex fashion.

It should be abundantly clear that the field of turbulence has no shortage of problems begging for solution, so let us see what strategy we might take to find a means of solving some of them. The first approach, (and perhaps the smartest) would be not to try. A lot of frustrating scientific careers have been spent probing the mysteries of turbulence to no avail. For those incautious souls who are not easily deterred by these forboding signs of peril, there are a few different alternatives one may pursue.

The first is to try direct numerical simulations of turbulent flows. One solves the Navier-Stokes equations exactly for a number of realizations of the flow field and then averages the results to obtain the various mean properties of interest. This approach is severely limited by the number of grid points one would need to solve a finite difference approximation to the Navier-Stokes equations at high Reynolds number. The number of storage locations

one would require has been estimated to be $\sim 10^{12}$, which far
exceeds the capability of the largest existing computers.

A modification of this approach is called subgrid modeling,
in which one computes only the large scale motion using a coarse
grid, and attempts to account for the influence of the small
(subgrid) scale motion on the large scale motion by suitable approximations. Unfortunately, the mathematical difficulties in
estimating the affect of the small scale motion on the large scale
motion are almost identical to those involved in approximating
the Reynolds stress which occur in the equation for the mean
velocity field. Thus, it is doubtful that there is any advantage
to subgrid modeling over trying to solve the equations for the
mean velocity field directly.

Another approach to solving turbulent flow problems is the
method of Reynolds stress modeling which is currently being actively
investigated by many researchers, prominent among them being Launder
in Great Britain and Lumley in the U.S. This approach is to use
a closed set of equations which govern the evolution of the Reynolds
stresses, instead of directly approximating them through the use of
an eddy diffusivity model such as that of Boussinesq, Taylor,
Prandtl and von Karman.

By manipulating the Navier-Stokes equations, one may derive
for the evolution of the stress $\overline{u_i' u_j'}$:

$$\frac{\partial}{\partial t} \overline{u_i' u_j'} + U_k \frac{\partial}{\partial x_k} \overline{u_i' u_j'} = P_{ij} + T_{ij} - D_{ij} - J_{ijk,k} \quad (1)$$

where

$$P_{ij} = - \frac{\partial U_j}{\partial x_k} \overline{u_i' u_k'} - \frac{\partial U_i}{\partial x_k} \overline{u_j' u_k'}$$

represents the production of $\overline{u_i' u_j'}$:

$$T_{ij} = p\overline{\left(\frac{\partial u_i'}{\partial x_j} + \frac{\partial u_j'}{\partial x_i}\right)}$$

accounts for the transfer between different components of the Reynolds stress;

$$D_{ij} = 2\nu \overline{\frac{\partial u_i'}{\partial x_k} \frac{\partial u_j'}{\partial x_k}}$$

accounts for its dissipation by viscosity; and

$$J_{ijk} = \overline{pu_i'}\,\delta_{jk} + \overline{\partial pu_j'}\,\delta_{ik} + \overline{u_i' u_j' u_k'} - \nu \frac{\partial}{\partial x_k} \overline{u_i' u_j'}$$

represents its diffusion by various means.

The evaluation of the left side of equation (1) is a simple matter, but the right side contains terms which are very difficult to evaluate. This method of Reynolds stress equation modeling is concerned with finding justifiable representations of the quantities, P_{ij}, T_{ij}, D_{ij} and J_{ijk} as functionals of the Reynolds stresses. There has been no general consensus as to the best approach to take in making these approximations, and each group of workers follow their own inclinations. It is too early to predict whether this approach will be successful in the long run.

In practice the equations contained in (1) are supplemented by an additional, heuristically conceived equation for the quantity $D = D_{ii}$, which is a dissipation term. One makes up an equation for describing what is known about the phenomenonology of the evolution of the dissipation, D. This is, in fact, the same principle which is used in the last approach to solving turbulent flows that we will discuss — that of phenomenonological modeling.

The eddy diffusivity model for approximating the Reynolds stress $\overline{u_i' v_j'}$ may be considered to be an elementary example of phenomenonological modeling. One makes a supposition as to how turbulent momentum transport occurs, and then models this process using velocity and length scales which are appropriate to the mechanism. For example, if we assume that in a turbulent jet, the transport of momentum occurs by the chaotic motion of eddys of a size and velocity characteristic of the jet itself, then one may hypothesize that $\overline{u'v'} = -\nu_T \, \partial U/\partial y$ and $\nu_T = b \, \Delta U \, L$ where L is the width of the jet, ΔU is the difference between the maximum velocity and that of the external flow and b is a constant to be determined empirically. One finds from experiments that if $b = .05 \rightarrow .1$ excellent results may be obtained.

To account for general turbulent flows one must be prepared to use phenomenonological modeling of a much more ambitious sort, i.e. one postulates the existence of equations for the evolution of a characteristic length and/or time and/or velocity scale of the turbulent motion. Each of these equations incorporates features of the known dynamics of the behavior of the appropriate scale. To tie these equations in with the averaged Navier-Stokes equations the practice has been to postulate the existence of the general constituitive relation:

$$-\overline{u_i' u_j'} = E\left(\frac{\partial U_i}{\partial x_j} + \frac{\partial U_j}{\partial x_i}\right) - \frac{q^2}{3} \delta_{ij} \qquad (2)$$

where $q^2 = \overline{u_i' u_i'}$ and E is a parameter dependent on the various scales whose development is being followed by the model equation

In 1942, Kolmogorov, in a paper which has only been recently discovered by the West, proposed a complete phenomenonological model, using (2) with $E = Aq^2/2w$, where A is a constant, q^2 is the turbulent kinetic energy and w is a inverse time scale, i.e. vorticity or frequency scale. q^2 and w are to be determined from the model equations:

$$\frac{Dw}{Dt} = -\frac{7}{10} w^2 + A' \frac{\partial}{\partial x_j}\left(\frac{q^2}{2w} \frac{\partial q^2}{\partial x_j}\right)$$

$$\frac{Dq^2}{Dt} = E\left(\frac{\partial U_i}{\partial x_j} + \frac{\partial U_j}{\partial x_i}\right)^2 - \frac{w}{2} q^2 + A'' \frac{\partial}{\partial x_j}\left(\frac{q^2}{2w} \frac{\partial q^2}{\partial x_j}\right) \quad .$$

A remarkably similar phenomenonological approach was suggested independently of Kolmogorov by Saffman in 1970. It also rely on equation (2), for this theory $E = Ae/w$ where e is a pseudo-energy. The model equations governing the development of e and w are

$$\frac{Dw^2}{Dt} = \alpha'w^2[2U_{i,j}^2 + 2(1-\eta)S_{ij}^2]^{1/2} - \beta'w^3 + A' \frac{\partial}{\partial x_j}\left(\frac{e}{w} \frac{w^2}{\partial x_j}\right)$$

$$\frac{D\ell}{Dt} = \underbrace{\alpha'' e[2S_{ij}^2]^{1/2}}_{\text{production}} \underbrace{- \ell w}_{\text{dissipation}} + \underbrace{A'' \frac{\partial}{\partial x_j}\left(\frac{\ell}{w} \frac{\partial \ell}{\partial x_j}\right)}_{\text{diffusion}} \Biggr\} \quad (3)$$

$$S_{ij} = \frac{1}{2}\left(\frac{\partial U_i}{\partial x_j} + \frac{\partial U_j}{\partial x_i}\right) \quad .$$

α', η, β', A, A', and A'' are universal constants which are determined by requiring the system of equations (3) to predict correctly some of the simple, empirically determined, properties of turbulent flows. For example it is known that the energy of homogeneous turbulence decays as the -6/5 power of the time, and if the equations in (3) are specialized to the case of homogeneous turbulence they predict that $\ell \sim t^{-2/\beta'}$. Therefore we may set $\beta' = 5/3$ to insure that the equations (3) are consistent with this decay law.

Another interesting feature of equations (3) is that they are designed to account for the presence of sharp turbulent — laminar interfaces which occurs in many turbulent flows, e.g. the jet. Within the turbulent zone $e > 0$, and in the laminar side, $e = 0$. It will be noticed that the diffusion term in these equations tends to zero as $e \to 0$. Since these equations, without the diffusion term admit the presence of discontinuities it is possible that the sharp boundary separating the turbulent and laminar flow regimes may be accounted for by these equations.

The system of equations (3) have been applied to a number of different turbulent flows and have achieved a modicum of success. For simple, self similar turbulent flows such as the jet, mixing layer, and wake, where the governing partial differential equations may be converted to o.d.e.'s, this phenomenonological theory has accurately reproduced the known empirical facts about these flows. In applications of this theory to a tubulent mixing layer between fluids of different densities at different Mach numbers, good agreement between this theory and experiments has been found, and, in addition, a prediction has been made for a problem which has not yet been studied experimentally. It is of great importance to see if this prediction turns out to be correct, because a truly successful theory of turbulence must be able to make accurate predictions, instead of just reproducing the results of experiments which have already been made, [i.e. "postdictions"].

In summary, the phenomenonological approach towards computing turbulent flows has had some initial success, but must undergo much more stringent tests in the future, e.g. by applying it to realistic complex flows, before one can judge whether it is truly successful or not. If this model continues to work, then one must search for a justification for it from the Navier-Stokes equations.

LECTURE V

FRACTALS AND TURBULENCE: ATTRACTORS AND DISPERSION

Benoit B. Mandelbrot

The renewal of interest in the mathematical aspects of turbulence has several independent and near simultaneous sources. The
dynamics approach well represented in this seminar is rooted in
the combined arguments of Lorenz 1963 and Ruelle & Takens 1970.
A separate approach started with the combined arguments of Kolmogorov 1962, Berger & Mandelbrot 1963, and Novikov & Stewart 1964;
the most recent statement is found in Mandelbrot 1976 and in the
book Fractals, Mandelbrot 1977.

The two approaches are bound to converge, if only because
both--and also those exemplified by the work of U. Frisch--make
vital use of nonstandard sets I have termed fractals. One notion
that is or should be stressed particularly in this kind of work
is the Hausdorff-Besicovitch dimension D. Since this notion is
classical but, so to say, somewhat obscure, it will be defined
and motivated below. However, it may be useful to say immediately that (in Fractals) a fractal set is defined as being such that

Hausdorff Besicovitch dimension > topological dimension.

For the standard sets of Euclid, on the contrary, these dimensions coincide. The term fractal structure may be defined loosely as synonymous with structure involving D. This quantity
becomes known as the fractal dimension.

The prototypical fractal is the Cantor set, and the product
of a Cantor set by an interval is also a fractal. This last

example enters in the well-known 1967 paper by Smale (see also
Lecture III above) and in Ruelle & Takens 1970. Each stage of
the Smale construction contracts the intercept of a torus into
N>1 domains contained in it, with the usual illustration assuming
N=2. In a different guise, contraction with N>1 also underlies
the processes due to Hoyle and to Novikov & Stewart which
Fractals describes under the name of "curdling". (The presently
available physical motivations are sketchy in both cases.) Curdling also involves a second (weakly motivated) assumption, which
has a counterpart in the the theory of contraction as restated in
Smale's lecture above (but not in the original). The assumption
is that each iteration replaces a set (either a curd or the
meridian intercept of a domain) by N subsets that are similar to
the original in a known ratio r<1.

The assumption concerning N is topological, but the assumption concerning r is metric in character. One metric property to
which it points is the fractal dimension, which we shall see is
given in this context by D=logN/log(1/r). There are many ways of
estimating the D from data (see Fractals) and their practical
importance suggests that the dynamics approach ought to be developed beyond topology, to include the fractal aspects. The same
remark applies to studies à la Lorenz 1963; there is no doubt
(though the fact remains to be proved) that the corresponding
"worse than strange" attractor is fractal; but its dimension is
not known to me. To evaluate it would be of intrinsic interest
and might help assess quantitatively rather than qualitatively to
what extent natural turbulence is modeled by simplified systems
of this kind (e.g., Hénon's model). (The value of D may play the
role occasionally played by the exponent in the spectral density.
It seems sometimes that simplified dynamic systems cease to be
meant to derive the Kolmogorov $k^{-5/3}$ spectrum, the quality of a
simplified system being judged on its ability to predict the -5/3
exponent.)

Thus, the term strange attractor used in Ruelle & Takens 1970
may well be a victim of the very success of the underlying approach, a more positively descriptive term becoming desirable.

One may suggest _fractal_ _attractor_. (One could go so far as to argue that the first words in the title of this talk are descriptive of the whole object of this seminar; however, this is not a suggestion I want to promote.)

    Two _aspects_ _of_ _the_ _notion_ _of_ _dimension:_ _motivation._ The mathematical characteristic which·the Lorenz and other "strange" attractors seem to share with the sets used in _Fractals_ is the following: It is known of the latter and suspected of the former that two alternative definitions of the notion of "dimension" yield distinct numerical values. The first is the topological dimension D_T. The second is the dimension D defined by Hausdorff and Besicovitch. Before we recall its definition, it is good to motivate D through the related concept of _similarity_ _dimension_ illustrated on Figure 1. (However, said illustration can be skipped; it is a variant of many in _Fractals_.) Figure 1 is the composite of two very-many-sided polygons one may call _teragons_. In Greek, _teras_ = a wonder or a monster, and in the metric system _tera_ = 10^{12}. One of these teragons is violently folded upon itself, being an advanced stage of the construction of a _plane-filling_ curve. By way of contrast, the second curve can be called a _wrapping_. Both are constructed by a von Koch cascade, from (a) an initial polygon, and (b) a standard polygon. The first construction stage replaces each side of the initial polygon by a rescaled and displaced version of the standard polygon. Then a second stage repeats the same construction with the polygon obtained at the first stage, and so on ad infinitum.

    The early construction stages are illustrated in Figure 2. The initial polygons are, respectively, a unit square and an irregular open polygon with N=17 sides. (It goes through every vertex of a certain lattice that is contained in the square.) Then each side of this 17-polygon is replaced by an image of its whole reduced in the ratio of $r=1/\sqrt{17}$. The result fills almost uniformly the shape obtained by replacing each side of the square by a certain polygon made of N=7 sides of length $r=1/\sqrt{17}$.

FIGURE 1

FIGURE 2

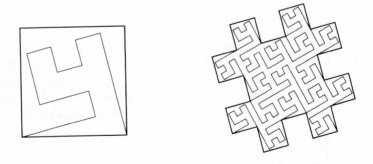

Incidentally, the familiar Peano curve and its variants drawn circa 1900 fill a square or a triangle, but recent Peano curves, like the present one, tend to involve more imaginative boundaries.

Since each construction stage multiplies length by a fixed factor $Nr>1$, both limit curves are of infinite length. But the filling tends to infinity more rapidly than its wrapping. This is expressed mathematically by the notion of similarity dimension. An intuitive explanation uses the following elementary fact: for every integer γ, the "whole" made up of a D dimensional parallelepiped may be paved by $N=\gamma^D$ "parts" which are parallelepipeds deduced by a similarity of ratio $r(N)=1/\gamma$. Hence, $D=\log N/\log(1/r)$. A dimension thus expressed as an exponent of self similarity continues to have formal meaning whenever the whole may be split up into N parts deducible from it by similarity of ratio r (followed by displacement or by symmetry). Such is the case with both limits here. For the wrapping, $N=7$ and $r=1/\sqrt{17}$, hence $D=\log 7/\log\sqrt{17}=1.3736$. For the filling, $N=17$ and $r=1/\sqrt{17}$, hence $D=\log 17/\log\sqrt{17}=2$. Thus the impression that the filling is more infinite than its wrapping is quantified by the inequality between their dimensions. The fact that the filling really fills a plane domain is confirmed by its dimension being $D=2$.

Hausdorff Besicovitch dimension and fractals. The first step in a general definition of D is to define the Hausdorff d-measure. Given a set S in a metric space and $\rho>0$, one covers S by balls with radii $\rho_m\leq\rho$, and one forms the sum $\Sigma\rho_m{}^d$; one takes the infimum of this sum over all coverings that satisfy $\rho_m\leq\rho$, then the limit of the infimum for $\rho\to 0$. The resulting $m_d(S)$ is by definition the Hausdorff d-measure of S. There exists a value of d, to be denoted by D, such that when $d>D$, $m_d(S)=0$ and when $d<D$, $m_d(S)=\infty$. This D is by definition the Hausdorff Besicovitch dimension.

Clearly, D is a metric rather than a topological property; I describe it as being a "fractal" property. More precisely, by a theorem of Szpilrajn (Hurewicz & Wallman 1941, p. 107), the topological dimension D_T and the above D are related by $D \geq D_T$. This explains the definition of fractals through $D > D_T$. The wrapping in Figure 1 is a curve of topological dimension 1; hence it is a fractal curve. For the triadic Cantor set, $D_T = 0$ while $D = \log 2 / \log 3$; hence it is a fractal. For the Cantor set considered by Smale in Lecture III above, $D = -\log 2 / \log \epsilon_2 < 1$. (However, by making N larger, one could obtain any $D < 2$ in this fashion.) In the case of homogeneous Kolmogorov turbulence in the Gaussian approximation, the isosurfaces of scalars satisfy $D_T = 2$ and $D = 8/3$, hence they are fractal surfaces. (While this value of D is used extensively in Fractals, a complete formal proof became available too late to be included in the bibliography; the reference is Adler 1977.)

Very frequently, D coincides with the similarity dimension examined in the preceding section.

Formal relation between fractal dimension and entropy-information. By theorems of Besicovitch and Eggleston (see Billingsley 1965), the fractal dimension frequently takes the form of an entropy-information; for example, there often exist an integer C and a discrete probability p_m such that $D = -\Sigma p_m \log_C p_m$. Example: for the triadic Cantor set, $p_1 = 1/2$, $p_2 = 0$, and $p_3 = 1/2$, while $C = 3$ (hence $C = \gamma = 1/r$); thus $D = \log_3 2$. The corresponding topological entropy comes to mind: it is equal to $\log 2$ with an unspecified basis for the logarithm. The metric entropy specifies this basis as being $1/r$.

The preceding formal relation may help bring the topological dynamic aspects and the fractal aspects of turbulence together.

Dispersion of a fluid line or tube. It is not for fractals and fractal dimension to provide ready-made theories, but they often help formulate empirical observations into geometric conjectures that suggest further experiments and mathematical problems. For turbulence, consider dispersion starting with a smooth curve such as a straight segment. According to one theory,

homogeneous turbulence causes the length to increase exponentially in time. It is easier to visualize the effects of a single
"pinch" of turbulent energy left to decay. Within the Richardso-
nian view of self similar turbulence, one can argue that said
effects subdivide into a sequence of stages, each of which multiplies the curve's length by some factor, either a fixed one or a
random one with a fixed distribution. This picture is a variant
of the Koch cascade of Figures 1 and 2. If it could be carried
out ad infinitum (neglecting the viscosity cutoff and the effects
of molecular diffusion), it would involve a fractal limit, and
the following alternative emerges: is this limit space-filling,
so that D=3, or such that D<3?

In planar reduction, D<3 corresponds to a curve like the
wrapping in Figure 1, while D=3 corresponds to the filling. Let
us first explore this second alternative. It views each 17-sided
polygon in the Koch construction of the filling as an eddy
(involving a net overall transport of matter). Observe that two
intervals of the initial curve having equal lengths are mapped on
two domains having equal areas. (In space: equal lengths map on
equal volumes.) However, this interesting complication is avoid-
ed if the Koch cascade does not initiate on the bottom side of a
square but on a domain. This domain can be taken to be the whole
square and the first cascade stage can be assumed to replace this
square by 17 squares collectively bounded by the first stage of
the wrapping, and so on. In this fashion, our curve-to-domain
application is embedded into a domain-to-domain application and
the image of a curve by the first application is identical to the
image of a domain by the second application. Any other initial
set again yields the same image if it is included in the square
and includes its bottom side. For example, one can represent a
fluid tube by the bottom 1/10-th of the original square, rounded
off to be shaped like one half of a sausage link. Our cascade of
transformations will make it into 17 smaller links, then 17^2,
etc. The limit will again be the whole interior of the wrapping
on Figure 1. Each stage of the mapping is discontinuous along
the lines where the preceding stage's link has been "pinched".

The (conjectural) turbulent mixing thus illustrated is grossly nonstationary. It is completely different from the usual stationary mappings such as the baker's transformation. Second difference: this "turbulent" mixing involves fixed points in exponentially increasing number and the baker's mixing has one fixed point. Third difference: there are reasons to expect the successive stages in this kind of turbulent cascade to proceed increasingly rapidly and the limit to be attained in finite time. (Denoting this time by t*, the length will vary like $(t*-t)^{-\alpha}$ with α a constant.)

The preceding model is readily generalized. Assuming that overall laminar motions are added, the final shape is no longer "globular", rather a long narrow strip, its overall shape being ruled both by turbulence and the laminar flow, and its detailed structure ruled only by turbulence. Furthermore, eddies can be made to be of different sizes, as for example in Figure 3. In comparison with the Koch method, the algorithm used here involves some complications, on which we shall not dwell. Broadly speaking, the initial shape ("sausage link") is a triangle. An earlier stage of the construction is visible (in reduced scale) in the eight triangles near the top of the picture, to the right.

An alternative conjectural view of turbulent dispersion involves the transformation of a smooth curve with $D=1$ into a fractal curve with $D<3$. The corresponding planar reduction is easily expressed; it would transform $D=1$ into $D<2$, as exemplified by the bottom fourth of the wrapping on Figure 1. However, the spatial form of this conjecture is hard to visualize. It is easier to imagine a spatial domain bounded by a surface of dimension $D=2$ being dispersed into a domain bounded by a surface of dimension $D\epsilon]2,3[$. See, for example, Fractals, p. 52, where (somewhat weak) reasons are given to believe that $D=8/3$. Similarly, one can work with a fluid filament with a diameter smaller than the original outer scale L and greater than the viscous inner scale η. As the cascade progresses and energy splits into eddies of decreasing diameter, this filament is taken to stretch and fold on itself. When the filament and eddy diameters are

FIGURE 3

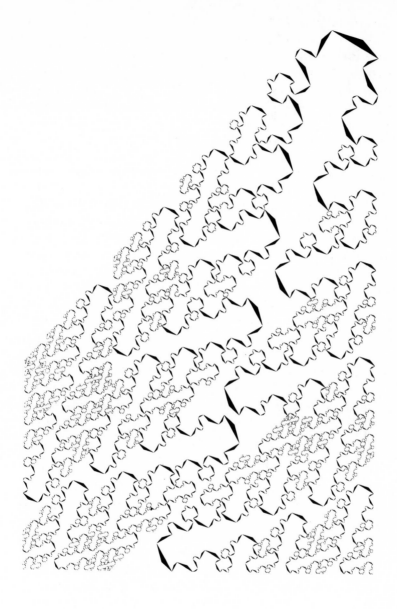

assumed proportional, we are led back to D=3. In order to achieve D<3, the eddy diameter must decrease more rapidly than the filament diameter. Such eddies will stretch the filament until it and the eddy have equal diameters. Thereafter, eddies will only affect the detail of the filament's surface; the filament's effective length will cease to change.

Of course, the preceding argument remains to be randomized, possibly along lines suggested by Robert Kraichnan. However, the feasibility of significant randomization depends strongly upon the dichotomy between D=3 and D<3. In the latter case, there is much room for it. In the former case, there is very little.

Now let us go from discrete pinches of turbulent energy on to homogeneous turbulence, approximating its effects to those of a sequence of pinches. In the case D<3, the filament's length will increase exponentially in time, as postulated by the theory to which we referred at the start. The sequence of successive transformations affecting it will be stationary.

Before attempting to model turbulent dispersion in detail, it may be advisable to analyze the evidence again, better than I could do here, to determine which of the above listed possibilities—or a farther variant—represents it properly. Once their task of helping sort alternatives is performed, the above cascade arguments should cease to be taken seriously, but the geometry they involve is likely to remain applicable.

REFERENCES

Adler, R.J. 1977. Hausdorff dimension and Gaussian fields. _The Annals of Probability_, Vol. 5, 145-151.

Berger, J.M. & Mandelbrot, B.B. 1963. A new model for the clustering of errors on telephone circuits. _IBM Journal of Research and Development_, Vol. 7, 224-236.

Billingsley, P. 1967. _Ergodic Theory and Information._ New York: Wiley.

Hurewicz, W. & Wallman, H.   1941.   Dimension Theory.   Princeton
   University Press.

Kolmogorov, A.N.   1962.   A refinement of previous hypotheses
   concerning the local structures of turbulence in a viscous
   incompressible fluid at high Reynolds number.   Journal of
   Fluid Mechanics, Vol. 13, 82-85.

Lorenz, E.N.   1963.   Deterministic nonperiodic flow.   Journal of
   the Atmospheric Sciences, Vol. 20, 130-141.

Mandelbrot, B.B.   1975.   Les objets fractals: forme, hasard et
   dimension.   Paris and Montreal: Flammarion.

Mandelbrot, B.B.   1976.   Intermittent turbulence and fractal
   dimension: kurtosis and the spectral exponent 5/3+B.   In Turbulence and Navier Stokes Equations, 121-145.   Ed. R. Temam.
   Lecture Notes in Mathematics 565.   New York: Springer.

Mandelbrot, B.B.   1977.   Fractals: form, chance, and dimension.
   San Francisco: W.H. Freeman and Company.

Novikov, E.A. & Stewart, R.W.   1964.   Intermittency of turbulence
   and the spectrum of fluctuations of energy dissipation.   (in
   Russian)   Isvestia Akademii Nauk SSR; Seria Geofizicheskaia,
   Vol. 3, 408-413.

Ruelle, D. & Takens, F.   1971.   On the nature of turbulence.
   Communications on Mathematical Physics, Vol. 20, 167-192 &
   Vol. 23, 343-344.

LECTURE VII

THE STRUCTURE OF LORENZ ATTRACTORS

R. F. Williams

The system of equations

$$(L) \begin{cases} \dot{x} = -10x + 10y \\ \dot{y} = 28x - y - xz \\ \dot{z} = -\frac{8}{3} z + xy \end{cases}$$

of E.N. Lorenz [L] has attracted much attention lately, in part because
of its supposed relation to turbulence. As was mentioned in Lecture I,
it is obtained by the truncation of the Navier-Stokes Equation. How
this system really relates to turbulence is not known yet and--as we saw

in earlier talks--there are many pros and cons for this. One thing is sure: this system certainly gives rise to a type of attractor which is nonclassical and not even strange in the sense of Smale (see Lecture III). Notice that according to the bifurcation classification of Lecture I, the system above gives rise to the "standard" Lorenz attractor and the goal of this lecture is to study it.

Our first task is to define the concept of "attractor" for a flow, a definition which was continuously omitted and not without reason. (The concept of attractor was defined in Lecture III but only for Axiom A diffeomorphisms and using the Spectral Decomposition Theorem of Smale.) From the physical viewpoint, it seems, that anything that is observable-- e.g., with non-zero probability--is associated in some way or another to something that "attracts" and therefore called "attractor." We shall use the following definition which seems to be accepted today by most who work in dynamical systems (see for example [BR], [S], [W2]). We start with the definition of some preliminary concepts.

Definition. Let M be a compact Riemannian manifold and $f_t: M \to M$, $t \in \mathbb{R}$ a smooth flow. A closed (f_t)-invariant set $\Lambda \subset M$ containing no fixed points is hyperbolic if the tangent bundle restricted to Λ, $T_\Lambda M$ can be written as a continuous Whitney sum

$$T_\Lambda M = E \oplus E^s \oplus E^u$$

with all three continuous subbundles (Tf_t)-invariant, E being the one-dimensional subbundle tangent to the flow, and there are constants c, $\lambda > 0$ such that

(i) $\|Tf_t(v)\| \leqslant ce^{-\lambda t}\|v\|$ for $v \in E^s$, $t \geqslant 0$

(ii) $\|Tf_{-t}(v)\| \leqslant ce^{-\lambda t}\|v\|$ for $v \in E^u$, $t \geqslant 0$.

Let now Λ be a closed (f_t)-invariant set. Consider the following properties:

(a) $f_t|\Lambda$ is a topologically transitive flow, i.e., $\exists\, x \in \Lambda$ such that $\{f_t(x)|t \in \mathbb{R}\}$ is dense in Λ (has a dense orbit);

(b) the periodic orbits of $f_t|\Lambda$ are dense in Λ ;

(c) there is an open set $U \supset \Lambda$ with $\Lambda = \bigcap\limits_{t\in\mathbb{R}} f_t U$

(d) U in (c) can be found such that $f_t(U) \subseteq U$

(e) Λ contains no fixed points and is hyperbolic.

If Λ satisfies: (b), (c), (d) it is called <u>attractor</u> (it should be noted here that most people would like to have (b) as a theorem, that is, one would make a weaker assumption (b′) and then show that (b′), (c), (d) imply (b));

(a), (b), (c), (e) it is called <u>basic hyperbolic set</u>;

(b) - (e) it is called <u>hyperbolic attractor</u>;

(a) - (e) it is called <u>basic hyperbolic attractor</u>.

Define for $x \in \Lambda$, Λ a hyperbolic set:

$$W^{ss}(x) = \{y \in M | \lim_{t \to \infty} d(f_t(x), f_t(y)) = 0\}$$

$$W^{uu}(x) = \{y \in M | \lim_{t \to \infty} d(f_{-t}(x), f_{-t}(y)) = 0\}$$

$$W_\varepsilon^{ss}(x) = \{y \in W^{ss}(x) | d(f_t(x), f_t(y)) \leqslant \varepsilon\}$$

$$W_\varepsilon^{uu}(x) = \{y \in W^u(x) | d(f_{-t}(x), f_{-t}(y)) \leqslant \varepsilon\}$$

where d is the metric induced by the Riemannian structure on M.
$W^{ss}(x)$ (respectively $W^{ss}_\varepsilon(x)$, $W^{uu}(x)$, $W^{uu}_\varepsilon(x)$) is called the
strong stable manifold (respectively the strong local stable, strong
unstable, strong local unstable manifold) at $x \in M$.

   Strong Stable Manifold Theorem (Smale, Hirsch, Palis, Pugh, Shub).
Let Λ be a hyperbolic set for a C^r flow (f_t) $(r \geqslant 1)$. Then for
$x \in \Lambda$, $f_t(W^{ss}(x)) = W^{ss}(f_t(x))$, $f_t(W^{uu}(x)) = W^{uu}(f_t(x))$. For
$\varepsilon > 0$ small

   ($1°$) $W^{ss}_\varepsilon(x)$, $W^{uu}_\varepsilon(x)$ are C^r-immersed discs with $T_x W^{ss}_\varepsilon(x) = E^s_x$,
$T_x W^{uu}_\varepsilon(x) = E^u_x$;

   ($2°$) $d(f_t(x), f_t(y)) \leqslant$ constant $e^{-\lambda t} d(x, y)$, for all $y \in W^{ss}_\varepsilon(x)$, $t \geqslant 0$
      $d(f_{-t}(x), f_{-t}(y)) \leqslant$ constant $e^{-\lambda t} d(x, y)$, for all $y \in W^{uu}_\varepsilon(x)$, $t \geqslant 0$
where λ is the constant from the hyperbolicity condition of Λ;

   ($3°$) $W^{ss}_\varepsilon(x)$, $W^{uu}_\varepsilon(x)$ vary continuously with x;

   ($4°$) If Λ is a hyperbolic attractor $\underset{x \in \Lambda}{\cup} W^{ss}_\varepsilon(x) = W^{ss}_\varepsilon(\Lambda)$ is a
neighborhood of Λ.

   Remark also that $W^{ss}(x) = \underset{t \geqslant 0}{\cup} f_{-t}(W^{ss}(f_t(x)))$,

$W^{uu}(x) = \underset{t \geqslant 0}{\cup} f_t(W^{uu}(f_{-t}(x)))$ for $x \in \Lambda$.

   The above theorem says essentially that we can choose a closed
neighborhood N of the hyperbolic attractor Λ in such a way that
$N \cap W^{ss}(x)$ will foliate N along Λ. In N define then the equiva-
lence relation $z_1 \sim z_2$ if and only if z_1, z_2 belong to the same
connected component of $W^{ss}(x) \cap N$. Denote by $q: N \to N/\sim$ the canonical "collapsing" map.

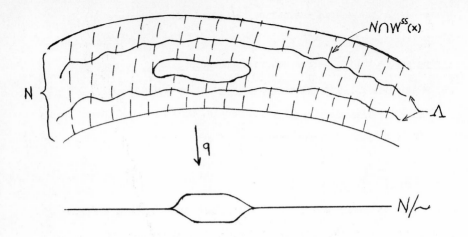

$N/\sim$ will be then in general a branched manifold. We shall abstain

here from exact definitions and refer the interested reader to [W2].

The above picture should be suggestive enough what a branched manifold

is.

Recall now that we have a flow on the manifold M ; Λ is invari-

ant by definition, but N is not. Actually N is invariant only un-

der the semiflow $(f_t)_{t \geqslant 0}$. Define then $g_t \colon N/\sim \to N/\sim$ for $t \geqslant 0$ by

$g_t(q(y)) = q(f_t(y))$ and notice that the definition is correct in view

of the preceding theorem. In this way we obtain a semiflow $(g_t)_{t \geqslant 0}$ on

the branched manifold $N/\sim$ which mirrors the behavior of our initial

flow "around" the hyperbolic attractor Λ . There are several advan-

tages to work with attractors under this form. First, they arise natu-

rally and lower the dimension. Second, they are real and give hence a

a strong intuition on the behavior of the flow. After collapsing, the

Lorenz attractor mentioned in Lecture I is the following branched mani-

fold:

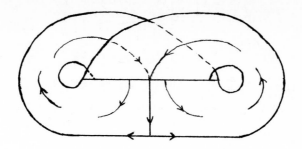

a picture present already in Lorenz' paper [L]. Compare the above picture with the one on page 15 of Lecture I to understand how the "collapsing" is done: the local stable manifolds lie perpendicular on this
branched manifold. A computer gives the same picture up to a smooth
deformation when programmed to find the attractor of the system (L).

So far, we reduced our study of an attractor to the study of a
branched manifold with a semiflow on it. Now another canonical construction demands to be done: given a point on our branched manifold κ
we would like to know the set of all its possible "prehistories" under
the semiflow $(g_t)_{t \geqslant 0}$. In this way we are led to consider the set

$$\hat{\kappa} = \{(k_t)_{t \leqslant 0} \mid k_t \in \kappa, \ g_s(k_t) = k_{t+s}, \ t + s \leqslant 0\} \ ,$$

which we recognize to be the inverse limit of the system $(X_t, \ u_{st})_{t \leqslant 0}$
where $X_t = \kappa$ for all t and $u_{st} : X_t \to X_s$ for $|t| \geqslant |s|$, $u_{st}(k_t)$
$= g_{s-t}(k_t) = k_s$, i.e., $\hat{\kappa} = \varprojlim_{t \leqslant 0} (X_t, \ u_{ts})$. Then for $z \geqslant 0$,

g_z induces naturally a map $\hat{g}_z = \varprojlim_{t \leqslant 0} g_z$ since $g_z \circ u_{st} = u_{st} \circ g_z$;

more precisely $\hat{g}_z((k_t)_{t \leqslant 0}) = (g_z(k_t)_{t \leqslant 0} = (k_{t+z})_{t+z \leqslant 0}$. The most

pleasant feature of this map is that it can be inverted, namely

$\hat{g}_{-z}((k_t)_{t \leqslant 0}) = (k_{t-z})_{t \leqslant 0}$, $z \geqslant 0$ and in this way $\{\hat{g}_z | z \in \mathbb{R}\}$ be-

comes a flow on $\hat{\kappa}$.

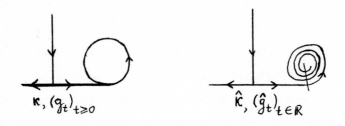

$$\kappa, (g_t)_{t \geqslant 0} \qquad\qquad \hat{\kappa}, (\hat{g}_t)_{t \in \mathbb{R}}$$

Theorem. Let Λ be a hyperbolic attractor of the flow (f_t) on

M , $\dim(W^{ss}(x) \cap \Lambda) = 0$ for all $x \in \Lambda$, N a closed neighborhood of

Λ foliated along Λ by local strong stable manifolds, $\kappa = N/\sim$ and

$(g_t)_{t \geqslant 0}$ the semiflow on κ induced by (f_t) . The following diagram

commutes

$$
\begin{array}{ccc}
\Lambda & \xrightarrow{\ f_z | \Lambda\ } & \Lambda \\
h \Big\downarrow & & \Big\downarrow h \\
\hat{\kappa} & \xrightarrow{\ \hat{g}_z\ } & \hat{\kappa}
\end{array}
$$

with the vertical arrows h homeomorphisms, $h(x) = (qf_t(x))_{t \leqslant 0}$,

$q: N \to N/\sim = \kappa$ being the collapsing map.

Proof. Continuity and closeness of h are trivial (by compacity).

The very definition of $\hat{\kappa}$ shows that each of its elements is of the

form $(q(x_t))_{t \leq 0}$ with $g_s(q(x_t)) = q(x_{t+s}) = q(f_s(x_t))$ for $s \geq 0$

and $t + s \leq 0$, hence $f_s(x_t) = x_{t+s}$, or $f_{-s}(x_{t+s}) = x_t$ for all

$t \leq 0$, $s \geq 0$, $t + s \leq 0$. Put here $t = -s$ and obtain $f_t(x_0) =$

$= x_t$ for all $t \leq 0$, which shows that any element in $\hat{\kappa}$ is of the

form $(q(f_t(x)))_{t \leq 0}$ and hence h is surjective. So far we did not

use the condition $\dim(W^{ss}(x) \cap \Lambda) = 0$ for all $x \in \Lambda$. This condi-

tion proves the injectivity of h. It is sufficient since we could

have situations in which $W^{ss}(x) \cap N$ has as dense subset $\Lambda \cap W^{ss}(x)$,

a pathological case excluded by our present assumption.   Q.E.D.

This shows that up to a canonical homeomorphism the behavior of

the flow on our hyperbolic attractor Λ is described by the inverse

limit $\hat{\kappa}$ with the flow $(\hat{g}_z)_{z \in \mathbb{R}}$

As a general principle, helpful in understanding the inverse limit

and its flow of a branched manifold, we can state:

    - if the "prehistory" of a point is unique, there is no

"doubling";

    - if there are lots of "prehistories," then there are lots of

"doublings" in the process of taking $\varprojlim$.

In what follows we want to apply this machinery to the study of

the Lorenz attractor. We saw already that as branched manifold, the

Lorenz attractor is:

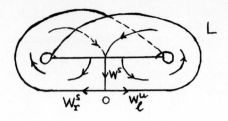

It should be emphasized here that L is actually canonically embedded in $\mathbb{R}^3$. The same is true for $\hat{L}$, that is, though our derivation of $\hat{L}$ from L is abstract, the end product $\hat{L}$ is embeddable in $\mathbb{R}^3$ and (with the gap between these constructions and the original equations discussed later in this talk) is the same as the Lorenz attractor, up to a topological (maybe smooth) conjugacy (see the theorem above and notice that the hypotheses are easily verified).

A main contribution to the understanding of the Lorenz attractor was done by Guckenheimer who recognized the basic geometric properties of the Lorenz attractor and proved the following theorem (see [G]).

Theorem (Guckenheimer). There are two topologically distinct types of Lorenz attractors.

Sketch of proof. To see what these attractors are, we shall trace the behavior of the unstable manifold of 0, W^u, which we split in two parts: the right unstable manifold W^u_r and the left unstable manifold W^u_ℓ; W^u_r enters the branch line at its rightmost point and W^ℓ_u enters first at its leftmost point (see picture above). The distinguishing feature of W^u is that it contains the only two semiorbits with a unique prehistory, namely the boundary of L . Now there are two possibilities: either W^u eventually hits W^s , or it does not. In the first case, topologically $\hat{W}^u \cup \hat{0}$ is a "figure eight", in the second case it is a line. Now we only have to remark that any homeomorphism $\phi: \hat{L} \to \hat{L}'$ will necessarily send $\hat{W}^u$ into $\hat{W}'^u$. This is proven by showing that each point $\hat{x} \in \hat{W}^u_\ell \cup \hat{W}^u_r$ lies in the interior of an interval $I' \subset \hat{W}^u_\ell \cup \hat{W}^u_r$ which is the "spine of a Cantor book" (i.e. $F \times I'$ where F is the cone over a Cantor set), whereas no other point of $\hat{L}$, with the possible exception of $\hat{0}$, lies in such a set. (For details see [W1])

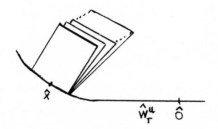

To extend this result, we must consider the quotient of L by its orthogonal trajectories. Denote the two generators

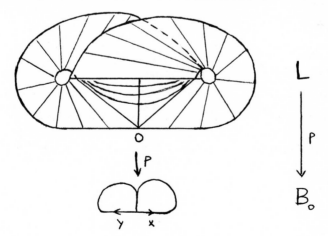

of the fundamental group of B_0 by x and y. Define a <u>kneading</u> <u>sequence</u> k as the pair (k_r, k_ℓ) where $k_r = p_0(W_r^u)$ thought of as a (possibly infinite) word in x and y and similarly for k_ℓ ; note that $k \in \Pi_1(B_0, 0)$.

Guckenheimer's result can now be rephrased in the following way: the two distinct topological types of Lorenz attractors are given by the cases: k_r, k_ℓ both finite and k_r, k_ℓ both infinite.

Given Λ a periodic orbit of the semiflow on L , $p(\Lambda)$ can be thought of as a positive word $w(\Lambda) \in \pi_1(B_0, 0)$ determined up to cyclic permutation. Define

$$\eta(x, y) = \sum_\Lambda \sum_\Gamma \frac{\Gamma w(\Lambda)}{\text{length}(w(\Lambda))}$$

where the sum is over all closed orbits Λ and for each Λ the second sum is taken over all cyclic permutaitons Γ of the word $w(\Lambda)$. Here retracing an orbit Λ is allowed, but this will produce a periodic word which thus will have fewer permutaitons.

We shall show below that the defintiion of η is actually inspired by the ζ-function of the first return map f on the branch line. Recall that the zeta-function of f is by defintion

$$\zeta_f(t) = \exp\left(\sum_{n=1}^\infty \frac{N_n}{n} t^n \right)$$

where N_n is the cardinality of the fixed point set of f^n . We shall prove now that

$$\exp \eta(t, t) = \zeta_f(t)$$

Indeed, if z is in the fixed point set of f^n and has minimal period p, then necessarily $n = pq$. Denote by Λ the closed orbit through z; it determines a word $w(\Lambda)$ of length p. Retrace it q times and then its contribution in the sum will be $\dfrac{pw(\Lambda)}{pq} = \dfrac{pw(\Lambda)}{n}$ since the number of possible cyclic permutations is only p, all other cyclic permutations being duplicates, so they don't count by our convention. Evaluate this at $(x, y) = (t, t)$ and get $\dfrac{p}{n} t^n$. But in the definition of ζ_f the closed orbit Λ of z counts as p points, namely the p points in which Λ intersects the branch line. Taking now the sum over all closed orbits Λ in both cases and rewriting the resulting sum after n we get the desired result.

We shall now show how η can be computed. For this purpose, denote by $F = p^{-1}(p(0))$, i.e. F is the orthogonal trajectory through 0 formed by two line segments. Denote by r_i the ith point where W_ℓ^u hits F; remark that $r_0 = \ell_0 = 0 \in F$. Define now a space of symbols $\sum$ in the following way:

(1) $[0, 1], [1, 0] \in \sum$

(2) $[i, j] \in \sum$ if and only if ℓ_i, r_i lie on the same side of 0

(3) if $[i, j] \in \sum$ and

    (a) $\ell_{i+1} = 0$, then $[0, j + 1] \in \sum$

    (b) $r_{j+1} = 0$, then $[i + 1, 0] \in \sum$

    (c) $\ell_{i+1} < 0 < r_{j+1}$, then $[i + 1, 0], [0, j + 1] \in \sum$

Order $\sum$ lexicographically and define the (in general) infinite matrix

$$A_{\sigma\tau} = \begin{cases} x \;, & \text{if } \sigma = [i, *] \;, \quad \tau = [i + 1, *] \;, \quad i = 0, 1, \ldots \\ y \;, & \text{if } \sigma = [*, j] \;, \quad \tau = [*, j + 1] \;, \quad j = 0, 1, \ldots \\ 0 \;, & \text{otherwise.} \end{cases}$$

where $\sigma, \tau \in \sum$; $*$ here means that anything can be substituted for it.

A straightforward formal computation shows that

$$\eta(x, y) = \sum_{n=1}^{\infty} \frac{\text{trace } A^n}{n} \in \mathbb{Z}[[x, y]] \;.$$

How intimately our kneading sequences are related to the function η is shown in the following.

<u>Theorem</u>. If (L, f_t) , (L', f_t') determine two Lorenz attractors, then $k = k'$ if and only if $\eta = \eta'$.

The main result ties up the inverse limit $\hat{L}$ with the kneading sequence:

<u>Theorem</u>. $\hat{L}$ is homeomorphic to $\hat{L}'$ if and only if $k = k'$.

In view of our previous theorem, this is equivalent to $\eta = \eta'$. We want to stress here that in the above theorem <u>any</u> homeomorphism is allowed; we <u>do</u> <u>not</u> suppose that it preserves the flows. We shall give here only the main steps of the proof and refer the interested reader to [W1].

- Step 1. Any homeomorphism h between $\hat{L}$ and $\hat{L}'$ maps $\hat{W}^u$ homeomorphically onto $\hat{W}'^u$. Deform h so that $h(\hat{0}) = \hat{0}'$. Use a barycentric approximation theorem and pass to the fundamental group to get a diagram

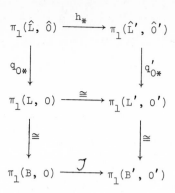

where $q_0(\overset{\Lambda}{x}) = x_0$, the projection on the 0th coordinate in the inverse limit. The diagram defines the map $\mathcal{J}$.

   - Step 2.   It is shown that $\mathcal{J}$ is either $(x, y) \mapsto (x, y)$, or $(x, y) \mapsto (y, x)$. This is the most tedious part of the proof which has to be itself split up into five subcases. First it is shown that $\mathcal{J}(x)$ is either entirely positive or entirely negative as a word in x and y . $\mathcal{J}$ must be an isomorphism so that after abelianisation, $\mathcal{J}$ becomes a matrix $\pm \begin{pmatrix} a & -c \\ b & -d \end{pmatrix}$ with a, b, c, $d \in \mathbb{N}$ and $ad - bc = \pm 1$. From here a careful analysis involving the cell complex structure of $\hat{F}$ proves our claim.

   - Step 3.   Using the relation $w(h(\Lambda)) = \mathcal{J}(w(\Lambda))$, Λ being a closed orbit, and Step 2, one shows that $\eta'(x, y) = \eta(x, y)$, or $\eta'(x, y) = \eta(y, x)$.

   -Step 4.   By our previous theorem and Step 3 we have: either $(k'_r, k'_\ell) = (k_r, k_\ell)$, or $(k'_r, k'_\ell) = (k_\ell, k_r)$.

   The cell-structure of $\hat{F}$ enables us to conclude:

   Theorem.   The periodic orbits are dense in each Lorenz attractor $\hat{L}$.

As mentioned before, $\hat{L}$ is homeomorphic to $\hat{L}'$, $k = k'$ and $\eta = \eta'$ are equivalent statements. It is tempting to conjecture that $k = k'$ if and only if $\eta_a = \eta'_a$, where the index a indicates that the η-function has been abelianized. The following counterexample shows that this is false.

Take $\alpha = xxxyxyx$, $\beta = yyxxyxx$ and $k = (\alpha xy, \beta xy)$, $k' = (\alpha yx, \beta yx)$, where x, y are as before the generators of the fundamental group of B. Clearly $k \neq k'$; we shall show, however, that $\eta_a = \eta'_a$.

The first thing to prove is that k and k' are indeed kneading sequences, that is the following two axioms are satisfied:

1) $k_\ell < k_r$, where $k = (k_r, k_\ell)$.

2) $k_\ell < w < k_r$, for any terminal word w of either k_ℓ or k_r.

The ordering in these two axioms has to be understood in the following way: x is the smallest symbol, y is the biggest, and the empty symbol $\emptyset$ is inbetween. So, for example $xx < x < xy$. Note also that the kneading sequences defined earlier for the Lorenz attractor always satisfy these two axioms. A glance at k and k' shows that both satisfy axioms 1 and 2 above.

This ordering enables us to construct the orbits defined by k and k', by ordering from left to right the branch line F, letting the orbit which starts at 1 follow the leaves of the Lorenz att actor and putting all members in the ordering corresponding to x's on the left and those corresponding to the y's on the right. For example, for k, the following numbers on F are obtained

$$\alpha xy \; : \; \begin{array}{ccccccccc} 1 & 4 & 8 & 13 & 7 & 12 & 5 & 9 & 14 \\ x & x & x & y & x & y & x & x & y \end{array}$$

$$\beta xy \; : \; \begin{array}{ccccccccc} 15 & 11 & 3 & 6 & 10 & 2 & 5 & 9 & 14 \\ y & y & x & x & y & x & x & x & y \end{array}$$

The matrix A in the formula for η will be then a 15×15 matrix.

Notice now that if w is a finite word, w will contribute to η if and only if

$$k_\ell < t(ww\ldots) < k_r \qquad\qquad (*)$$

t indicating "terminal sequences."

Write now k and k' respecting the ordering in the following way:

$$\underbrace{\alpha xy \; \Delta \; \overbrace{\alpha yx \quad \text{same} \quad \beta xy \; \Delta' \beta y}^{k'} \; x}_{k}$$

By Δ we mean $\{w | w$ finite satisfying (*) and $\alpha xy < ww\ldots < \alpha yx\}$. The gap labeled "same" consists of words w which occur both in η and η' . Δ' is defined similarly to Δ . Thus to show $\eta_a = \eta_a'$ we need to establish a one-to-one correspondence between Δ and Δ' so that corresponding words are the same when abelianized.

We proceed as follows: any word $w \in \Delta$ must begin with α and since (*) involves all terminal sequences of $ww\ldots$, any word $w \in \Delta$ must end in y, so is of the form αvy. Similarly, $w' \in \Delta'$

has the form $\beta v'x$. Thus the first candidates are $w = \alpha y$ $w' = \beta x$, and these are indeed the only words of length 8 in Δ, Δ'. Now,

$$(\alpha y)_a = x^5 y^3 = (\beta x)_a.$$

Next let v_i and v'_i be the v's and v''s of length i from above. One finds $v_1 = \emptyset = v'_1$, $v_2 = \{xy\}$, $v'_2 = \{yx\}$ and $(\alpha xyy)_a = (\beta yxx)_a$. Finally, one can check that $v_i = v'_i$ for i = 3, 4, 5, 6, 7. Thus the (abelianized) words of η_a and η'_a of length $\leqslant 15$ are the same. Using now the formula of the η-function

$$\eta(x,y) = \sum_{n=1}^{\infty} \frac{\text{trace } A^n}{n}$$

for our case (the matrices A and A' for η and η' are 15 × 15), it follows that trace A^i = trace A'^i for i $\leqslant$ 15 when abelianized and hence for all i. We conclude that $\eta_a = \eta'_a$.

For completeness we list the last 5 cases of $v_i = v'_i$: $\{yx^2\}$, $\{yx^2y\}$,$\{xy^2x^2, yx^3y, yx^2yx\}$, $\{yx^2yx^2, yx^2yxy\}$ and $\{xy^2x^3y, yx^3yxy, yx^2yx^2y, yx^2yxyx, yx^2y^2x^2\}$.

We stop here our qualitative analysis of the Lorenz attractor and end with some remarks concerning the gap between Lorenz' equations (L) and our previous exposition. As was already remarked in Lecture I, most of the analysis aimed to pin down the attractor is done by computer work; there are many unproven things, of which the most important is: show rigorously that the attractor defined by (L) has the form given in the picture on page 14 of Lecture I (i.e., is after collapsing along the strong stable manifold our branched manifold L).

A second important thing to be proven is the following: Show that there exists a field of strong stable directions, in a certain sense "normal" to $\hat{L}$. In other words, prove the existence of the foliation by strong stable manifolds.

Finally we formulate two conjectures. As mentioned before, the periodic orbits are dense in $\hat{L}$. I think that they:

(a) link one another, and

(b) are knotted.

History

Several years ago, Jim Yorke figured out some things about the Lorenz equation and got other mathematicians interested. He gave some talks on the subject, including one here at Berkeley. Ruelle, Lanford and Guckenheimer became interested and did some work on these equations. Unfortunately, except for the recent preprint of Ruelle [R], Guckenheimer's paper [G] is the only thing these four people ever wrote on the subject--as far as I know.

BIBLIOGRAPHY

[B]   Bowen, R., Equilibrium States and the Ergodic Theory of Anosov
            Diffeomorphisms, Springer Lecture Notes in Mathematics,
            470.

[G]   Guckenheimer, J., A Strange, Strange Attractor in Marsden,
            McCracken, The Hopf Bifurcation and Applications,
            Applied Mathematical Sciences 19, Springer Verlag,
            1976.

[HP]   Hirsch, M., Pugh, C., The Stable Manifold Theorem, Global Anal-
            ysis, Proceedings Symposia in Pure Mathematics, vol.
            14, American Math. Soc., 1970.

[HPS]   Hirsch, M., Pugh, C., Shub, M., Invariant Manifolds (to appear).

[HPPS] Hirsch, M., Palis, J., Pugh, C., Shub, M., Neighborhoods of
            Hyperbolic Sets, Inventiones Math. 9, 121-134 (1970).

[L]   Lorenz, E., N., Deterministic Nonperiodic Flow, Journal of At-
            mospheric Sciences, 20(1963), 130-141.

[R]   Ruelle, D., The Lorenz Attractor and The Problem of Turbulence,
            preprint from the conference on "Quantum Dynamics
            Models and Mathematics" in Bielefeld, September 1975.

[S]   Smale, S., Differentiable Dynamical Systems, BAMS, 13(1964),
            767-817.

[W1]   Williams, R., The Structure of Lorenz Attractors, preprint
            Northwestern University, 1976.

[W2]   Williams, R., Expanding Attractors, Publications Mathématiques
            iHES no. 43, (1974), 169-203.

[W3]   Williams, R., One Dimensional Non-wandering Sets, Topology,
            6(1967) 473-487.

[BR]   Bowen, R., Ruelle,. D., The Ergodic Theory of Axiom A Flows,

            Inventiones Math. 29, 181-202 (1975).

APPENDIX TO LECTURE VII: COMPUTER PICTURES OF THE LORENZ ATTRACTOR

Oscar Lanford

1. First fifty loops of one branch of the unstable manifold of the origin of the classical Lorenz attractor: r=28.

2. Next fifty loops of the same branch of the unstable manifold of the origin.

3. Lorenz attractor (one branch of the unstable manifold of the origin) for another parameter: r=40.

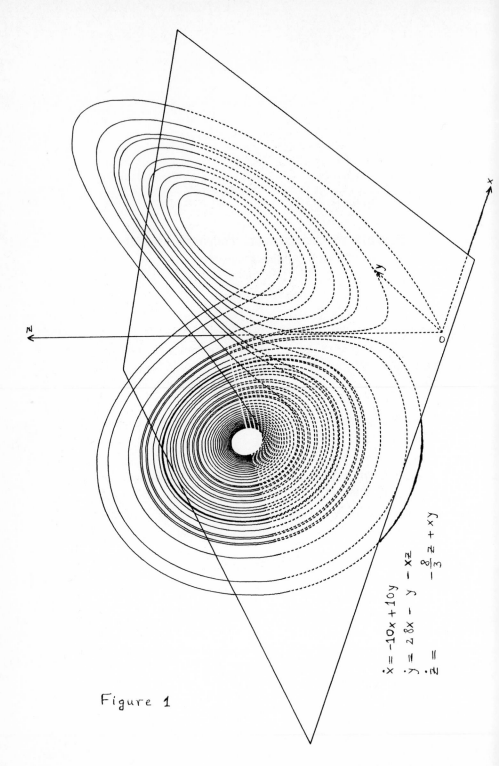

Figure 1

$$\dot{x} = -10x + 10y$$
$$\dot{y} = 28x - y - xz$$
$$\dot{z} = -\frac{8}{3}z + xy$$

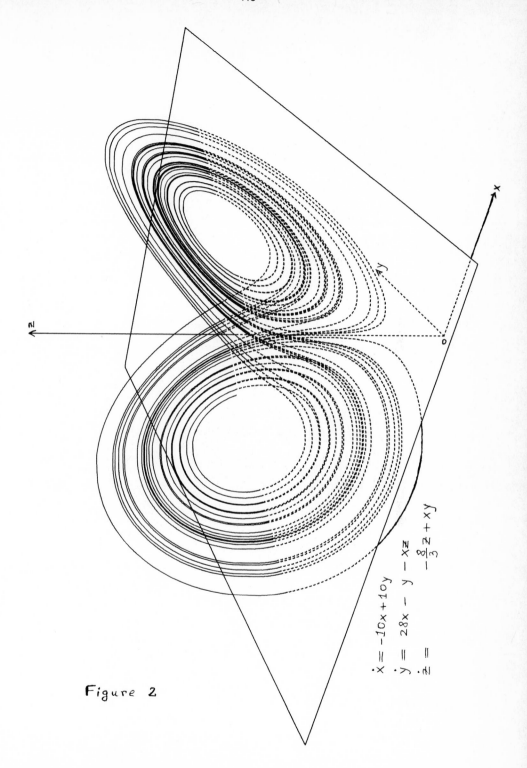

Figure 2

$$\dot{x} = -10x + 10y$$
$$\dot{y} = 28x - y - xz$$
$$\dot{z} = -\frac{8}{3}z + xy$$

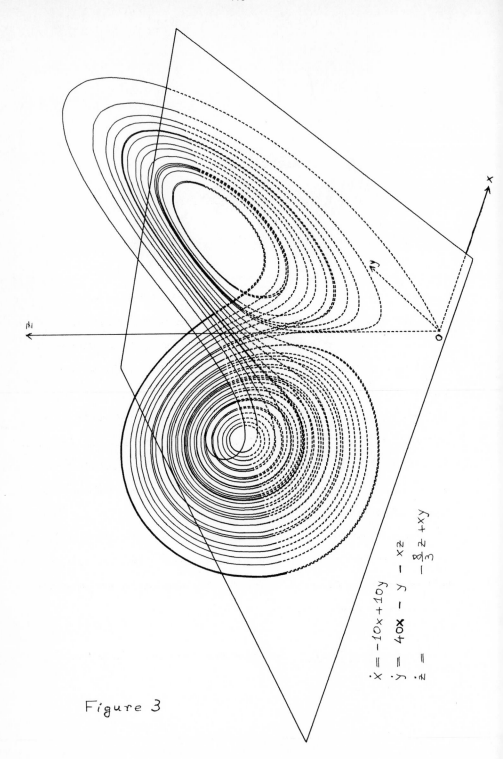

Figure 3

$$\dot{x} = -10x + 10y$$
$$\dot{y} = 40x - y - xz$$
$$\dot{z} = -\frac{8}{3}z + xy$$

LECTURE VIII

A MODEL FOR COUETTE FLOW DATA

Rufus Bowen

This talk will be about the statistical behavior of flows. The
first part will review some well-known examples which display different
types of behavior. With these examples in mind we will look at experimental data on Couette flow, giving a family of flows whose behavior
roughly mimics this data.

Let $\varphi_t : M \rightarrow M$ be a differentiable flow on a compact manifold M.
Differentiable functions $f: M \rightarrow \mathbb{R}$ are called observables. An experiment
consists of picking a point $x \in M$ (initial conditions) and plotting
$g(t) = f(\varphi_t x)$ as a function of t. One tries to extract physical reality
from these graphs.

For a moment let's think about an experiment whose outcome is "heads"
or "tails" at each of a countable number of times. There are many conceiv-
able mechanisms to do the "flipping" in this coin experiment. Here are
three examples of what could happen.

Experiment 1 (Trivial). Heads always comes up, no matter what.

Experiment 2 (Tame). Let S^1 be the unit circle in R^2 and let $\varphi: S^1 \to S^1$ be rotation by some angle α incommensurable with 2π. Define $f(z)$ by

$$f(z) = \begin{cases} \text{heads if } z \text{ above or on x-axis} \\ \text{tails if } z \text{ belox x-axis.} \end{cases}$$

One picks $z \in S^1$ and then observes

$$f(x), \ f(\varphi(z)), \ f(\varphi^2(z)), \dots .$$

It is a classical result that for any z the points $z, \varphi(z), \varphi^2(z), \dots$ are uniformly distributed on S^1. This translates into the following statement for our experiment:

$$\frac{\text{number of heads in } n \text{ trials}}{n} \to \frac{1}{2} \quad \text{as } n \to \infty$$

no matter what z is and the convergence is uniform.

Experiment 3 (Random). This is the usual flipping of a fair coin. All sequences of heads and tails are equally likely. For almost all experiments

$$\frac{\text{number of heads in } n \text{ trials}}{n} \to \frac{1}{2} \quad \text{as } n \to \infty$$

but certainly this is not true for all experiments and the convergence is not uniform.

In a little while we will look at differentiable flows displaying the above types of behavior. But first we give a definition.

Definition. A smooth flow $\varphi_t: M \to M$ is ergodic if there is a set $\hat{M} \subset M$ so that $M \setminus \hat{M}$ has Lebesgue measure 0 and for every (continuous) observable f the limit

$$\bar{f} = \lim_{T \to \infty} \frac{1}{T} \int_0^T f(\varphi_t x) \, dt$$

exists for $x \in \hat{M}$ and is independent of $x \in \hat{M}$. The flow is uniquely ergodic if $\hat{M} = M$. Finally, if U is an open subset of M, then one says φ_t is ergodic or uniquely ergodic on U if the above statements hold with M replaced by U.

Notice that all three of the coin experiments were ergodic but only the first two uniquely ergodic. The notion of ergodicity of course arose in early discussions of statistical mechanics and the above definition is certainly what physicists had in mind. Notice that no invariant measure μ is mentioned but one automatically gets one by defining

$$\int f \, d\mu = \bar{f} \ .$$

If φ_t is a conservative system, then there will be a smooth invariant measure m given by Liouville's theorem and φ_t will be ergodic in the sense of our definition precisely if m is an ergodic measure for φ_t (and then $\mu = m$). If φ_t is not a conservative system, φ_t can still be ergodic but one expects μ to be concentrated on a set of Lebesgue measure 0. We now review a few examples of flows.

    1. **Attracting fixed point.** Here $p \in M$ is fixed under φ_t (i.e. $\varphi_t x = x$ all t) and there is a neighborhood U of x so that $\lim_{t \to \infty} \varphi_t x = p$ for any $x \in U$. Then clearly φ_t is uniquely ergodic on U with $\bar{f} = f(p)$ and $\mu = \delta_p$.

    2. **Attracting closed orbit.** Here $\varphi_\tau p = p$ for some smallest positive $\tau > 0$ and there is a neighborhood U of the circle $\Lambda = \{\varphi_t p : t \in [0,\tau]\}$ so that $\varphi_t x \to \Lambda$ as $t \to \infty$ for all $x \in U$. One has unique erogicity on U with $\bar{f} = \frac{1}{\tau} \int_0^\tau f(\varphi_t p)\, dt$ and μ spread around evenly on Λ according to the time parameter.

    3. **Linear flow on the torus.** One thinks of the torus T^2 as the quotient group $T^2 = \mathbb{R}^2/\mathbb{Z}^2$ where $\mathbb{Z}^2 = \{(m,n) \in \mathbb{R}^2 : m,n \text{ are integers}\}$. Now fix a vector $\vec{v} = (v_1, v_2) \in R$ and define the flow φ_t on T^2 by

$$\varphi_t((x,y) + \mathbb{Z}^2) = (x + tv_1, y + tv_2) + \mathbb{Z}^2.$$

If either $v_i = 0$ or if v_1/v_2 is rational, then φ_t is a periodic flow -- i.e. for some $\tau > 0$ one has $\varphi_\tau p = p$ for all $p \in T^2$. If v_1/v_2 is irrational, then φ_t is ergodic and in fact uniquely ergodic. One sees the time averages $\bar{f}$ existing for all points of T^2 by using the fact that φ_t is almost periodic. In the irrational flow case that $\bar{f}$ is independent of initial condition follows by Fourier analysis. Taking a circle cross-section to φ_t the Poincaré map is just a rotation (experiment 2).

4. Horocycle Flow. The classical horocycle flow is obtained by letting φ_t be left multiplication by $\begin{pmatrix} 1 & t \\ 0 & 1 \end{pmatrix}$ on the compact homogeneous space $SL(2,R)/\Gamma$ (Γ a discrete group). Hedlund proved this was ergodic and Furstenberg [Fu] recently proved unique ergodicity. Marcus [M] has now verified unique ergodicity for a general class of horocycle flows. We mention this example not because it represents something for Couette flow but because it is on the borderline between tame and random experiments. By analogy with our model coin experiments, the attracting fixed point and closed orbit are trivial and the irrational flow on T^2 is tame. The horocycle flow is uniquely ergodic but graphs of observables $g(t)$ will not show the almost periodic behavior that occurs for the torus flow. The horocycle flow is tame in the sense of a gorilla locked in a cage.

5. Axiom A attractors. Here there is an open set U with $\varphi_t U \subset U$ for $t \geq 0$ and $\Lambda = \bigcap_{t>0} \varphi_t U$ an attractor, i.e. $\varphi_t x \to \Lambda$ as $t \to \infty$ for $x \in U$. There is a hyperbolicity assumption on the derivative of φ_t on Λ. We refer the reader to [Sm] for an exact definition and to Smale's lecture for an example. The set Λ could be M (Anosov case) or may resemble locally products of manifolds with Cantor sets. The flow is ergodic on U but not uniquely ergodic, and the measure μ lies on Λ. This is proved in [B-R], but there is a very large literature concerned with special cases of this and statistical properties of μ. Here are a few references: [An], [A-S], [A-W], [BO], [B-R], [O-W], [Ra 1], [Ra 2], [Ru 1], [Si 2]. The measure μ here is constructed by a thermodynamic formalism and it will

be interesting to see whether the various formal quantities (entropy, energy function, pressure) have a connection with the physical quantities with the same names for some real flow.

6. <u>Billiards</u>. Some convex smooth obstacles are fixed on a torus and a moving particle bounces off them, obeying the usual laws. Sinai [Si 1] showed this is ergodic.

7. <u>Lorenz attractor</u>. This example was thoroughly discussed in Williams' lecture. Modulo details of the type he mentioned, Lanford and Wong [W] have each proved results about maps of [0,1] which imply that the Lorenz attractor is ergodic.

By analogy with the coin experiments we classify the above flows as follows:

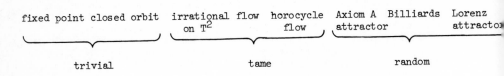

Ergodicity is the first step in the statistical study of φ_t. Once it is verified one looks for properties of the canonically defined invariant measure μ. Let's consider a few of these.

A. <u>Correlation Functions</u>. Fix observables f and g and define

$$\beta(t) = \overline{f}\,\overline{g} - \int f(x) g(\varphi_t x)\ d\mu(x)\ .$$

If $\lim\limits_{t\to\infty} \beta(t) = 0$ for any f, g, then φ_t is called <u>mixing</u>. The fixed point, horocycle flow, Axiom A attractors and billiards are mixing while the closed orbit, and torus flow are not. One doesn't know about the Lorenz attractor, but one would guess its behavior is like Axiom A attractors and billiards. It is not known how fast $\beta(t) \to 0$ for Axiom A attractors. Ratner and I guess that $\beta(t) \to 0$ faster than some $t^{-\alpha}(\alpha > 0)$ but not exponentially fast. Here that f and g are differentiable is vital.

B. <u>Central Limit Theorem</u>. One says φ_t satisfies the C.L.T. if for every (differentiable) observable f (with a few obvious exceptions) there is a constant $\sigma = \sigma(f) > 0$ so that

$$\lim_{T\to\infty} \mu\left\{x \in M: \frac{\int_0^T f(\varphi_t x)\,dt - T\bar{f}}{\sigma\sqrt{T}} \geq a\right\} = \frac{1}{\sqrt{2\pi}} \int_a^\infty e^{-u^2/2}\,du$$

Ratner [Ra 1] proved the C.L.T. for Axiom A flows and Bunimovic [Bu] for billiards. The point here is that fluctuations from the average obey a normal law in the limit of large time. It remains unknown whether in these examples the fluctuations obey the law of the iterated logarithm.

C. <u>Bernoulliness</u>. A mapping of a measure space is Bernoulli if it is measure theoretically isomorphic to the march of time for the (usual) flipping of an n-sided coin with specified probabilities for the sides. Ornstein [O] and his colleagues have a very good understanding of Bernoulli processes. Using these ideas one knows that billiards [G-O]

and Axiom A flows [Ra 2] are Bernoulli. A warning is in order: knowing
that a flow is Bernoulli does not give the CLT since differentiable
(i.e. nice) observables are not preserved by measure theoretic conjugacy.
The horocycle flow is not Bernoulli for entropy reasons; however, it is an
outstanding problem whether this flow is loose Bernoulli in the sense of
Feldman [Fe].

Now that we have seen some of the types of behavior flows can have
let's take a look at some experimental data. Fenstermacher, Swinney,
and Gollub [F-S-G], [G-S] made velocity measurements on Couette flow and
Fourier analyzed their data. Here is a table of some of their results
[F-S-G]

Observed Frequencies

R/R_c	ω_1	ω_2	ω_3	B	ω_3/ω_1
1	--	--	--	--	--
1.1	(a)	--	--	--	--
10	1.328	--	.82	--	.62
11	1.320	.1	.86	--	.65
17	1.310	0	.93	.44	.71
19.8	1.317	--	.94	.45	.71
22.4	1.34	--	--	.46	--

(a) present but not accurately measured.

Here R/R_c is a normalized Reynolds number which we think of as a parameter
for the system. As this

parameter is increased, the system passes through various types of flow
and this is reflected by the different frequencies observed in the
Fourier analysis of the velocity correlations (above). ω_1, ω_2 and ω_3
are sharp frequencies while B is a broad band of frequencies whose
center is given in the table.

In line with the Ruelle-Takens theory of turbulence [R-T], [Ru 2],
we will try to give a sequence of flows on a finite dimensional manifolds
M whose behavior matches the data in the table. The experimenter's observable is a function f: M → R. All our flows will be ergodic and so at each
stage there will be a canonical invariant measure μ. The correlation

$$\beta(\tau) = \int f(x) f(\phi_\tau x) d\mu(x)$$

therefore is defined and it satisfies

$$\beta(\tau) = \lim_{T \to \infty} \frac{1}{T} \int_0^T f(\phi_t x) f(\phi_{\tau+t} x) dt$$

for almost every initial condition x. The experimenter takes a large T
and calculates $\beta(\tau)$ by this formula (for $\tau < T$). He then finds the
Fourier transform of $\beta(\tau)$ (power spectrum) and the dominant frequencies
are recorded in the table. It is ergodicity therefore that provides the
justification of the table; the problem raised before on how fast $\beta(\tau) \to 0$
has to do with whether the Fourier transform is a function or a measure.

Initially the liquid is moving in a simple manner and our model flow is an attracting fixed point. With the appearance of the frequency ω_1 at $R/R_c \sim 1.1$ the flow becomes an attracting closed orbit with period ω_1^{-1} (Hopf bifurcation). Let D be a disk with dim D = dim M-1 which is transverse to this periodic orbit. Following a point $x \in D$ along the flow until it hits D again induces a Poincaré map $\psi: D \rightarrow D$; ψ is a contraction and its unique fixed point p is where D intersects the attracting closed orbit of the flow.

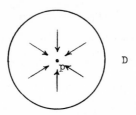

The next change is the appearance of a second frequency ω_3. The easiest model for this is a Hopf bifurcation of the map $\psi: D \rightarrow D$ into $\tilde{\psi}: D \rightarrow D$ which has an invariant circle K (i.e. $\tilde{\psi}k = k$) which is an attractor:

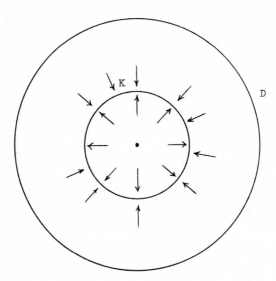

$\tilde{\psi}$ maps K onto itself. The image of K under the flow φ_t is an invariant torus $T^2 = \{\varphi_t y : y \in K, t \in \mathbb{R}\}$. This torus is an attractor for the flow φ_t.

As R/R_c changes this picture of an attracting torus persists for awhile. The circle K may change position slightly (which is unimportant) and the map $\tilde{\psi}: K \to K$ may change (which _is_ important). Each diffeomorphism of the circle has a rotation number; here this number is just ω_3/ω_1. If ω_3/ω_1 is irrational, then $\tilde{\psi}$ is just a rotation by the angle $(\omega_3/\omega_1)2\pi$ (in terms of some coordinates). If ω_3/ω_1 is rational, then $\tilde{\psi}$ could be a rotation or it could have the following property: there is a set S of N points such that $\tilde{\psi}(S) = S$ and $\tilde{\psi}^n(x) \to S$ as $n \to \infty$ for almost every $x \in K$. The orbits of S under the flow φ_t is then one or more circles which are attractors for φ_t. The integer N in this case must be such that $(\omega_3/\omega_1)N$ is an integer. In physical terms we are describing resonance as opposed to interference (of the two frequencies ω_1 and ω_3). Our intuition tells us that one will see this resonance mostly likely in cases where N (the denominator of ω_3/ω_1) is small. There are two "reasons" for this: when N is large one needs the behavior of many points (the set S) to match up just right and even when this does happen the set S starts to look pretty dense in K (i.e. it's hard to see $S \neq K$). Arnold [Ar] shows that for a general one parameter family of analytic maps of the circle one expects an interval of resonance about each rational rotation number with the size of the intervals decreasing fast enough as the denominator increases so that irrational rotation occurs with positive probability. This means that we expect to see the attracting torus persist for a while with ω_3/ω_1 varying and that when this picture breaks down ω_3/ω_1 will be rational with small denominator.

At this point we stumble onto the small frequency ω_2. I will not try to explain this in terms of a model flow φ_t on M. Small frequencies tend to be noise or numerical artifacts. It is tempting to view ω_2 as some topological noise in the flow φ_t which leads to the onset of the band B. However, in earlier experiments [G-S] ω_2 occurs before ω_3 does (hence its index) and disappears well before B appears. You can see why I want to ignore it.

Now we tackle the broad band B. This band corresponds to the random flows we mentioned before, as opposed to the trivial and tame flows representing 0, 1, and 2 frequencies. This band _experimentally_ appears at ω_3/ω_1 = .71 but its _real_ appearance is at $\omega_3/\omega_1 = 2/3$. Of course, I want this to be reality because 2/3 is the strongest candidate for a rational number > .62 which will be "seen," in terms of the earlier discussion. When the random type attractor (band) first appears qualitatively in the flow on M, there may be very little to distinguish it quantitatively from a periodic orbit. A topological change is not necessarily noticed by approximate values of a quantitative observation f: M → R. As the diameter of the attractor transverse to the flow increases, no topological change takes place but its presence becomes noticeable.

When the diffeomorphism of the circle $\tilde{\psi}: K \to K$ becomes resonant at rotation number 2/3, one would guess first at a picture like

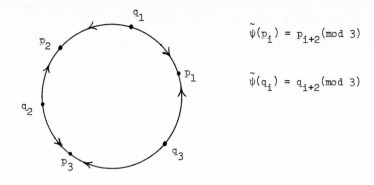

$$\tilde{\psi}(p_i) = p_{i+2} \pmod 3$$

$$\tilde{\psi}(q_i) = q_{i+2} \pmod 3$$

Here $S = \{p_1, p_2, p_3\}$ is an attracting periodic orbit and $\{q_1, q_2, q_3\}$ a repelling periodic orbit (you don't see it). The band effect means that each p_i is replaced by a neighborhood D_i

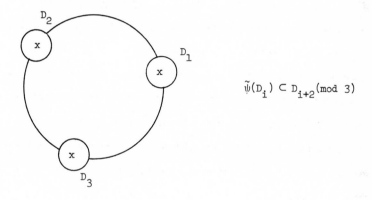

$$\tilde{\psi}(D_i) \subset D_{i+2} \pmod 3$$

and $\tilde{\psi}/D_i$ <u>not</u> just a contraction. Instead $\Lambda = \bigcap\limits_{n=0}^{\infty} \tilde{\psi}^n(D_1 \cup D_2 \cup D_3)$ is a complicated (random) attractor. One has many orbits in the attractor and so the frequencies form a band. When the D_i are small in diameter (when they first arise) the band should be narrow.

In the above picture one expects the frequency ω_3 to be replaced by a band situated near the old ω_3. Unfortunately this is contrary to the data. The data shows ω_3 persisting sharply for a little while and the band B centered about $\frac{1}{3}\omega_1$ $(= \frac{1}{2}\omega_3$ upon formation of B, $\sim\frac{1}{2}\omega_3$ a bit later). The simplest model we could think of for this behavior was as follows: A neighborhood of the circle K (in the cross-section to the flow φ_t) is foliated by circles whose length is twice as long as K. When resonance occurs, the attractiveness of K is nearing an end and a neighborhood of one of these nearby circles L is becoming attractive. The map near L behaves like the picture we gave above for K but rejected. $\tilde{\psi}$ moves points along L roughly as far as along K, but since L is about twice as long as K the frequencies in the band come out right.

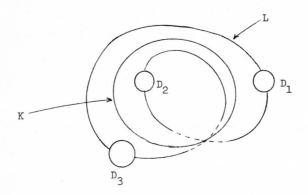

For a moment, K and the new attractor generated near L fight it out; K loses and so ω_3 disappears.

Let $\Lambda = \bigcap_{n=0}^{\infty} \tilde{\psi}^n(D_1 \cup D_2 \cup D_3)$ and Ω be the set of orbits of φ_t passing through Λ. Then Ω is a random type attractor for the flow (we have in mind an Axiom A attractor or a Lorenz type attractor, but there is no way of telling from the type of reasoning we are indulging ourselves in). The set Λ lies in a set D transversal to the flow.

The frequency ω_1 arose by the periodic return to the transversal D by an attracting closed orbit; ω_1^{-1} was the time it took to return. When this closed orbit changed to a torus the time to return at least slightly would depend on the point $x \in K$ one is at. However, because the flow on the torus is uniquely ergodic (tame) there is a well defined average return time that appears quickly for all orbits. This is why a sharp ω_1 persits. For the random attractor Ω there is still an average return time by ergodicity, but convergence does not take place for all orbits and fluctuations are very prevalent. The expanding diameters of the sets $\Lambda \subset D_i$ mean both that the band B will broaden and that there will be enough "visible" variation in return time (to $\Lambda \subset D$) so that ω_1 fades out. Notice that this fading out could be gradual and requires no topological change in the nature of the flow.

We have thus gone through a sequence of flows whose statistical behavior at least roughly matches the [F-S-G] data. Let us return to the most interesting and questionable part of the model-namely the existence of a foliation of a neighborhood of K by circles of twice the length. One way to get such a foliation is an follows: let $g: N \to N$ be a map of a disk with one fixed point p such that $g^2(x) = x$ for all $x \in N$. One

can suspend this to get a flow whose orbits are circles (g is the

Poincaré return map on the transversal section N); the circle through p

will be half as long as all the other circles. The most obvious reason

for a map like g to be lurking in the Couette flow would be an order 2

symmetry in the physical system. Upon asking Lanford about this, he

immediately suggested the symmetry between top and bottom of the cylinders.

Intuition tells us that the appearance of the band B at half the expected

frequency could in fact arise from an order 2 symmetry in various ways; one

should think of the circle foliation as a model idea and not a rigid des-

cription.

It should be clear that the above sequence of flows was put together

in an entirely speculative manner. These flows provide a simple model

which is consistent with the experimental data, but clearly I have made no

attempt to "prove" it is the "correct" model. One could argue that what I

have talked about today is worthless, because it seems that flows on finite

dimensional manifolds can model most any experimental data. The second

part of that thought is what I'd like you to remember from this lecture.

References

[A-W]   R. Adler and B. Weiss, Similarity of automorphisms of the torus, Memoirs AMS 98 (1970).

[An]   D. Anosov, Geodesic flows on closed Riemann manifolds with negative curvature, Proc. Steklov Inst. Math. 90 (1967).

[A-S]   D. Anosov and Ya. Sinai, Some smooth ergodic systems, Russ. Math. Surveys 22 (1967), no. 5.

[Ar]   Arnold, Small denominators I, mappings of the circle onto itself, Transl. AMS, ser. 2, 56 (1965), 213-284.

[B-R]   R. Bowen and D. Ruelle, The ergodic theory of Axiom A flows, Inventiones Math. 29 (1975), 181-202.

[Bo]   R. Bowen, Equilibrium states and the ergodic theory of Anosov diffeomorphisms, Lecture notes in math. 470, Springer-Verlag.

[Bu]   L. Bunimovich, Central limit theorem for a class of billiards, Th. Prob. and its Appl. 19 (1974), 65-85.

[Fe]   J. Feldman, New K-automorphisms and a problem of Kakutani, preprint.

[F-S-G]   R. Fenstermacher, H. Swinney and J. Gollub, Transition to turbulence in a rotating fluid, preprint.

[Fu]   H. Furstenberg, The unique ergodicity of the horocycle flow, Lecture Notes in math. 318, Springer-Verlag, 95-115.

[G-O]   G. Gallavotti and D. Ornstein, Billiards and Bernoulli Schemes, Comm. Math. Phys. 38 (1974), 83-101.

[G-S]   J. Gollub and H. Swinney, Onset of turbulence in a rotating fluid, Phys. Rev. Letters 35 (1975), 927-930.

[M]   B. Marcus, Unique ergodicity of the horocycle flow: variable negative curvature case, Israel J. Math. 21 (1975), 133-142.

[O]   D. Ornstein, Ergodic theory, randomness and dynamical systems, Yale Univ. Press.

[O-W]   D. Ornstein and B. Weiss, Geodesic flows are Bernoullian, Israel J. Math 14 (1973), 184-197.

[Ra 1]   M. Ratner, The central limit theorem for geodesic flows on n-dimensional manifolds of negative curvature, Israel J. Math. 16 (1973), 181-197.

[Ra 2]   M. Ratner, Anosov flows with Gibbs measures are also Bermoullian, Israel J. Math. 17 (1974), 380-391.

[Ru 1]   D. Ruelle, A measure associated with Axiom A attractors, Amer. J. Math. 98 (1976), 619-654.

[Ru 2]   D. Ruelle, The Lorenz attractor and the problem of turbulence, preprint.

[R-T]   D. Ruelle and F. Takens, On the nature of turbulence, Comm. Math. Phys. 20 (1971), 167-192.

[Si 1]   Ya. Sinai, Dynamical systems with elastic relfections. Russ. Math. Surveys 25 (1970), 137-189.

[Si 2]   Ya. Sinai, Gibbs measures in ergodic theory, Russ. Math. Surveys 166 (1972), 21-69.

[Sm]   S. Smale, Differentiable dynamical systems, Bull. AMS 73(1967), 747-817.

[W]   S. Wong, Some metric properties of piecewise monotonic mappings of the interval, preprint.

TWO ATTEMPTS AT MODELING TWO-DIMENSIONAL TURBULENCE

Harland M. Glaz

Two related problems of considerable interest in the theory of turbulence are (1) the explicit construction of dynamical systems with only a finite number of degrees of freedom which faithfully model Navier-Stokes turbulence, and (2) deciding exactly what is meant by the words "faithfully model." I.e., one must decide which physical observables (e.g., energy and vorticity spectra, velocity correlations) are to be studied and test any proposed model to see that the results of measurements of these observables in experiments (that is, in flows described by the full Navier-Stokes equations) manifest themselves in the model. In this talk, two models are presented and each is followed by a discussion of their possible merits. This approach may be contrasted with part of R. Bowen's talk in which observables (namely, a sequence of bifurcations) were noted from an experiment (see [GS1]) but the explicit construction of a model with the appropriate behavior was left open.

Most remarks of substance in the following will be for the case of 2-dimensional flows although similar discussions are possible in the 3-dimensional case.

Notation: An underline denotes a vector. N = number of dimensions of the flow. The spatial coordinates are indicated by $\underline{x} = (x,y,z)$, and the velocity and vorticity fields are denoted $\underline{u}$, $\underline{\xi} = \nabla \times \underline{u}$ respectively. Summation from 1 to N is indicated for repeated Greek subscripts. Finally, $\mathcal{D} \subseteq \mathbb{R}^N$ will be the region in which the fluid is confined.

The first system to be considered is the vortex model. To see why this model is important, one notes from experiment that the vorticity distribution is a critical observable in turbulence. In particular, there appears to be a tendency towards the formation of macroscopic (i.e., large-scale) vortices as the velocity field develops from its initial value according to the Navier-Stokes equations (see, e.g. [LT 1]). The vortex model, although a rather coarse idealization of real flows, is specifically designed to study this observable directly. This contrasts with our second model which requires a considerably more roundabout analysis when looking at such qualitative observables as vorticity structure.

To begin, we record some basic results. The Navier-Stokes equations are

$$
\begin{cases}
1) & \underline{u}_t + (\underline{u} \cdot \nabla)\underline{u} = -\text{grad } p + \nu \Delta \underline{u} + \underline{F} \\
2) & \text{div } \underline{u} = 0
\end{cases}
$$

where p = pressure, $\underline{F}$ = external force, ν = viscosity. $\underline{F}$ will always be taken to be zero and for the vortex model, $\nu = 0$ as well. Also, $\underline{u}$ is two-dimensional.

Since div $\underline{u} = 0$, there exists a stream function $\psi(\underline{x})$ such that

3) $\underline{u} = (-\partial\psi/\partial y, \partial\psi/\partial x)$.

By the definition of vorticity it follows that

4) $\Delta\psi = -\xi$

The first idea in the vortex model is the use of a point vortex. A single point vortex located at $\underline{x}_0 \in \mathcal{A}$ is defined by the vorticity field

5) $\xi(\underline{x} - \underline{x}_0) = K_0\delta(\underline{x} - \underline{x}_0)$

where $K_0 \equiv$ strength of the vortex $= \int_{\mathcal{A}} \xi d\underline{x}$. Note that the sign of K_0 indicates the orientation of the vortex. The stream function for a single point vortex is now easily obtained:

6) $\psi(\underline{x} - \underline{x}_0) = -\Delta^{-1}K_0\delta(\underline{x} - \underline{x}_0) = \dfrac{K_0}{2\pi} \log |\underline{x} - \underline{x}_0|$.

Suppose now that there are M of these point vortices located at the points $\underline{x}_j$, $j = 1, \ldots, M$, and that there are no other contributions to the vorticity field. Assume further that each vortex retains its structure in time, i.e., that they remain point vortices. Then, Kelvin's Theorem (the total vorticity of a fluid element is conserved) implies that each K_j remains constant. Since the vortices move with the fluid, the velocity of the jth vortex, denoted by $\underline{u}_j$, will simply be due to the velocity field generated by all of the other vortices (point vortices do not self-interact). That is,

7) $d\underline{x}_i/dt = \sum_{j \neq i} \underline{u}_j$, $i = 1, \ldots, M$.

This defines the vortex model. I.e., one looks at a random initial distribution of point vortices and studies their motion using (7).

The most important feature of the model is that it is a Hamiltonian system. To see this, define

8) $q_i = \sqrt{|K_i|}\, x_i$, $p_i = \sqrt{|K_i|}\, \mathrm{sgn}(K_i) y_i$

for $i = 1, \ldots, M$ and

9) $H(\underline{q}, \underline{p}) = -\frac{1}{2\pi} \sum_{i<j} K_i K_j \log r_{ij}$

where $\underline{q} = (q_1, \ldots, q_M)$, $\underline{p} = (p_1, \ldots, p_M)$, and $r_{ij} = |\underline{x}_i - \underline{x}_j|$. Using (7), it is easy to verify that

10) $dH/dt = 0$

11) $dq_i/dt = \partial H/\partial p_i$, $dp_i/dt = -\partial H/\partial q_i$, $\forall i$.

Together, (10) and (11) define a Hamiltonian system with "energy" H and canonical coordinates (q_i, p_i).

It should be remarked at this point that the system (7) - (11) has another possible interpretation in terms of 2-D plasma physics. If the x_j denote the positions of very long and thin rods aligned parallel to an external magnetic field directed perpendicular to the plane of motion which react via the Coulomb potential, then the rods remain so aligned, the Hamiltonian describes the motion of the rods according to a "guiding-center drift," and H may be interpreted as the total potential energy due to the Coulomb interactions. See [MJ1] for details. Actually, the form of H is somewhat more complicated if one takes boundaries into account. See [S1].

A subject of intense recent interest is the so-called "negative temperature state" phenomenon for the vortex model. This arises by taking M to be large and applying the methods of classical statistical mechanics to the Hamiltonian (9). For the details of the derivation, see [On1], [Ch1], [MJ1], [FO1], [Mon 1], and references in these papers. It turns out that if certain plausible, but unconvincing, assumptions are made, then vortices of the same orientation coalesce if and only if the thermodynamic temperature derived from statistical mechanics (see [Kh 1]) is negative. (Note: the reason that it is possible for the temperature to be negative is that $\mathcal{B}$ is taken to have finite volume which implies that phase space has finite volume as well.) Therefore, since temperature is a function of the total energy H, this implies that the system should form macroscopic vortices for some ranges of H and should tend not to form them for other values of H. This hypothesis is clearly testable, say with a computer. In [MJ 1] tests along this line are reported, the conclusion being that negative temperatures do indeed correspond to macroscopic vortex formation. However, the evidence is clearly far from conclusive.

Whatever the outcome of this argument, it is clear that the model is useful in studying the vorticity distribution for the special case of collections of point vortices. What is not at all clear is the question: can the model be used as an approximation for a real 2-D flow which initially is given by a collection of vortices, not necessarily point vortices, which do not necessarily retain their structure for all time? This question must be answered in the affirmative if the vortex model is to be useful in the theory of turbulence. Chorin's random vortex model (see [Ch 4]), viewed as an attempt to overcome some of the restrictions of the vortex model above, is important in this regard.

A final remark about the vortex model: in the 3-D case, one might consider a collection of long thin vortex tubes, take a surface perpendicular to each tube, and look at the motion of the tubes on the surface in analogy with the guiding-center plasma interpretation. Of course, in a truly 3-D flow, the surface will move with the fluid making it difficult to parametrize the surface for all time. See [Ch 1].

The second model to be discussed is the Fourier Mode model. This model is applicable to the Navier-Stokes equations in any number of dimensions and also to a one-dimensional model equation, Burger's equation:

12) $$u_t + uu_x = \nu u_{xx} \ .$$

From this point on, we shall assume that $\mathcal{B}$ is a box upon which periodic boundary conditions are imposed. For convenience, the side lengths of the box are taken to be $L = 2\pi$ – this makes the wave numbers integers. It is easy to show that changing L merely scales the unknowns in the resulting system below and does not affect its qualitative behavior.

Temporarily, we work explicitly with the pressure, and expand $\underline{u}$ and p into Fourier series,

13) $\underline{u}(\underline{x},t) = \sum_{\underline{k}} \underline{u}(\underline{k},t)e^{i\underline{k}\cdot\underline{x}}$, $p(\underline{x},t) = \sum_{\underline{k}} p(\underline{k},t)e^{i\underline{k}\cdot\underline{x}}$

and proceed as follows to obtain a system of O.D.E.'s –

(I) Substitute (13) into (1), (2) (or (12)).

(II) Use (2) to eliminate the $p(\underline{k},t)$ terms (analogously, in physical space one can show that $\nabla^2 p = -\nabla((\underline{u}\cdot\nabla)\underline{u})$).

(III) Equate the coefficients of $e^{i\underline{k}\cdot\underline{x}}$ for each $\underline{k}$ in the resulting expression.

The end result is an infinite set of ordinary differential equations in the infinite collection of variables $\{u_\alpha(\underline{k},t): \alpha = 1,\dots,N;$ $k_i = 0, \pm 1, \pm 2,\dots$ for $i = 1,\dots,N\}$. It reads

14) $\left[\dfrac{\partial}{\partial t} + \nu k^2\right] u_\alpha(\underline{k},t) = -\dfrac{i}{2} P_{\alpha\beta\gamma}(\underline{k}) \sum_{\underline{p}} u_\beta(\underline{p},t) u_\gamma(\underline{k}-\underline{p},t)$

15) $k_\alpha u_\alpha(\underline{k},t) = 0$

where

16) $P_{\alpha\beta\gamma}(\underline{k}) = k_\beta P_{\alpha\gamma}(\underline{k}) + k_\gamma P_{\alpha\beta}(\underline{k})$

17) $P_{\alpha\beta}(\underline{k}) = \delta_{\alpha\beta} - k_\alpha k_\beta/k^2$.

These equations are valid for $N = 2$ or 3. However, they simplify
somewhat for $N = 2$, and a much simpler result is available for $N = 1$.
A more complicated expression, but of the same form as (14) - (17) could
be obtained in the case $N > 3$. The derivation (straightforward) of
(14) - (17) is left to the reader. For more details, particularly in
regard to numerical implementation, see [Or 2]. Observe that $\underline{u}(\underline{x},t)$
real implies that

18) $\underline{u}(-\underline{k},t) = [\underline{u}(\underline{k},t)]^*$

The final step necessary to obtain a dynamical system (in the usual sense)
is

(IV) Truncate (14) - (15) in such a way as to respect the condition
(18). That is, set all except a finite number of Fourier modes equal to
zero for all time; but if $\underline{u}(\underline{k},t)$ is retained then so must $\underline{u}(-\underline{k},t)$. No

further restrictions will be placed on the truncation so in principle an infinite collection of dynamical systems can be obtained from (14) - (15) in the above manner. However, for large-scale simulations, one usually retains all $\underline{k}$ satisfying $|\underline{k}| < K$ where K is chosen so that Fast Fourier Transforms can be used to handle the convolution sums in (14). See [Or 2] for details.

We specialize now to the inviscid case and write down the system for the spherical truncation –

$$19) \quad \dot{u}_\alpha(\underline{k},t) = -\frac{i}{2} P_{\alpha\beta\gamma}(\underline{k}) \sum_{\substack{|\underline{p}|,|\underline{q}| < K \\ \underline{p}+\underline{q} = \underline{k}}} u_\beta(\underline{p},t) u_\gamma(\underline{q},t)$$

$$20) \quad k_\alpha u_\alpha(\underline{k},t) = 0$$

where there is one equation for each $|\underline{k}| < K$ and, in (19), one equation for each $\alpha = 1,\ldots,N$.

We digress for a moment and consider how well a truncation in the form (19)-(20) can be expected to approximate a real flow (1)-(2). There are two difficulties. First, in a real turbulent flow, ν is very small but certainly $\nu \neq 0$. A heuristic argument of Batchelor's ([Ba 12]) indicates that for a 3-D flow the $\nu \to 0$ limit is singular in the sense that a finite rate of energy dissipation persists in the limit whereas it will be shown below that energy is conserved for the inviscid system (19)-(20). The argument requires that all Fourier modes be present (i.e., no truncation) so if his result is true, it would argue against using (19)-(20) as a model for 3-D turbulence. However, for Burger's equation (see [Ho 1]) and for the 2-D system with periodic boundary conditions (see [EM 1]), it turns out that solutions of the Navier-Stokes equations with positive viscosity approach the solutions of the inviscid equations as $\nu \to 0$ and one has uniform convergence on compacta for the Fourier transforms of the solutions.

The second difficulty lies in the truncation--not all modes are retained. However, at least in principle, it should be possible to obtain an excellent approximation by keeping all modes with wave numbers less than that used to describe, say, the intermolecular distance. This is clearly impractical and the real question is whether or not a reasonable truncation is also a good approximation.

In any event, however these two problems may be resolved, it is reasonable to assume that the Fourier mode representation should give an accurate account of at least some aspects of turbulence. We now turn, after some preliminaries, to the question of locating observables of real flows in the model. The energy and vorticity spectra are directly built into the system. This aspect shall not be discussed in any detail here; instead we shall concentrate on the more qualitative observables, specifically the question of vorticity distribution for 2-D flows.

Even though the system (19)-(20) is not Hamiltonian (at least, it has not been shown that it is), a Liouville Theorem still holds. That is,

$$21) \qquad \sum_{|\underline{k}| < K} \frac{\partial}{\partial u_\alpha(\underline{k},t)} \left(\frac{du_\alpha(\underline{k},t)}{dt} \right) = 0$$

The proof is trivial. The significance of the result is that the dynamical system preserves Lebesgue measure in the phase space consisting of the dynamical variables $\{u_\alpha(\underline{k})\}$ (see, e.g., [Kh 1]).

Another important result which is also easy to prove is that energy is conserved--

$$22) \qquad \frac{d}{dt} \sum_{|\underline{k}| < K} u_\alpha(\underline{k},t)u_\alpha(-\underline{k},t) = 0.$$

If we restrict attention to the 2-D case only, there is another quadratic constant of the motion called the enstrophy or total vorticity --

$$23) \qquad \frac{d}{dt} \sum_{|\underline{k}| < K} k^2 u_\alpha(\underline{k},t)u_\alpha(-\underline{k},t) = 0.$$

Remark: In the untruncated system (still with $\nu = 0$), there are an infinite number of integrals of the motion, one for each moment of the vorticity distribution. These are all lost in the truncation except (23).

Taking advantage of the relations $k_\alpha u_\alpha(\underline{k}) = 0$, $u_\alpha(-\underline{k}) = u_\alpha(\underline{k})^*$, and $\dot{u}_\alpha(\underline{0}) = 0$, it is only necessary to retain the independent modes in any truncation. Say there are M such independent modes and name them $\underline{x} = (x_1,\ldots,x_M)$. Then, the system takes the form

$$24) \qquad \dot{\underline{x}} = \underline{F}(\underline{x})$$

where each component of $\underline{F}$ is purely quadratic in the x_i. The constants of the motion and Liouville's Theorem may be written

$$25) \qquad \sum_{i=1}^{M} \alpha_i x_i^2 = E, \qquad \sum_{i=1}^{M} \beta_i x_i^2 = \Omega, \quad \text{div}(\dot{\underline{x}}) = 0$$

where E = energy, Ω = enstrophy, and the α_i, β_i are easily determined constants for a particular truncation. Using this formulation, we proceed to discuss the dynamics of the Fourier mode system.

As has been shown, there are at least two constants of the motion, E and Ω. Therefore, the motion takes place on $S = S_E \cap S_\Omega$ where S_E = surface of constant energy E and S_Ω = surface of constant enstrophy Ω. Clearly dim $S = M-2$. If F_t denotes the time t map of the flow induced by $\underline{F}$, then $F_t : S \to S$ defines the dynamical system. It is important to note that there is an F_t-invariant measure (for all t) on S which is absolutely continuous with respect to $(M-2)$-dimensional Lebesgue measure. It is

$$26) \qquad d\mu = d\Sigma / \|\text{grad } E\| \ \|\text{grad } \Omega\| \sin \theta$$

where $d\Sigma = (M-2)$-dimensional Lebesgue measure and θ is the angle between grad E and grad Ω. The derivation (which is elementary) for the case in which energy is the only conserved quantity may be found in [Kh 1] and the derivation of (26) is an easy generalization.

Of course, (24) is an initial value problem. So, when considering a statement such as "macroscopic vortices form in 2-D flow," the natural translation into the Fourier mode model would read: "there exists an F_t-invariant set $X \subseteq S$ which may be identified with large-scale vortices in physical space and which is a global attractor for the flow." However, such a hypothesis is certainly untenable in the case where the viscosity $\nu = 0$ since the existence of an attractor (which attracts a set of positive measure) would violate the condition that $d\mu$ is F_t-invariant. In addition, we note that a precise characterization of the set X is by no means easy to come by.

One important way to overcome this problem is to introduce a small viscosity into the system. This eliminates conservation of measure in phase space, energy, and enstrophy. Also, $\underline{u} = 0$ becomes a global attractor. However, interesting phenomena may well occur between the initial data and the final decay, perhaps including "attractors" as defined above. Along the same lines, it is possible to reintroduce a stirring force $\underline{F}$ into (1) and into the model in such a way that E and Ω are conserved even though $\nu = 0$. In this way, one could study the so-called inertial range, i.e., those wave-numbers intermediate between the very high modes where viscosity is predominant and the very low ones which describe the larger scales of the motion. See [Ch 1] for details. Of course, such a model would have to be very large-scale (i.e., would involve retaining a large number of modes) to be realistic. Also, it is hard to see how $\underline{F}$ could be constructed so as to be time-independent. Hence, the F_t-invariant measure $d\mu$ would not exist. In any event, this may well be an advantage since the possibility of physically interesting attractors would reemerge.

Two remarks are in order concerning the last two paragraphs.
First, the discussion applies to any observable which is translatable
into the Fourier mode model. Second, the contrast between the $\nu > 0$
system and the $\nu = 0$ system illustrates the possible danger involved
in taking the $\nu \to 0$ limit in the Navier-Stokes equations.

Next, let us return to the $\nu = 0$ case. A paper of Chorin's
(see [Ch 2]) points out that there exists an invariant set of measure
zero for the untruncated system (19)-(20) as well as for the truncated
system, and elements of this set are the Fourier transforms of vortices
in physical space. Explicitly, the construction reads

$$27) \quad u_1(\underline{k},t) = -ik_2\psi(\underline{k},t), \quad u_2(\underline{k},t) = ik_1\psi(\underline{k},t)$$

$$28) \quad \psi(\underline{k},t) = C \exp(iak_1 + ibk_2)/H(\underline{k},t)$$

where ψ is the stream function and C, a, b are constants. Let $X \subseteq S$
denote this invariant set as before. Of course, $\mu(X) = 0$. However, one
may conjecture the following: there exists an F_t-invariant set Y
with $X \subseteq Y \subseteq S$ and $0 < \mu(Y) < \mu(S)^*$. This hypothesis is meant as a
replacement in the $\nu = 0$ case for the various possible attractor
hypotheses in the $\nu > 0$ case, and in this regard it has a defect. Namely,
how are we to regard the F_t-invariant set $S-Y$ which also satisfies
$\mu(S-Y) > 0$? First, one might consider the limit $M \to \infty$ where $M =$ number
of retained modes. Letting μ_M, S_M,Y_M correspond to μ, S, Y respectively

---

*(Note: See [Mo 1] for some interesting results concerning the existence
of Y given the invariant set X, in the case of some Hamiltonian
systems -- which is not the case here.)

it could be conjectured that $\mu_M(S_M - Y_M)$ becomes negligible while at the same time Y_M retains its physical significance. Second, we may take the point of view that $\underline{u}$ in (1), (2) is a _random_ field, i.e., that there exists a probability space $(\Omega, \mathcal{B}, \lambda)$ where Ω is identified as the state space and $\underline{u}(\underline{x}) = \underline{u}(\underline{x}, \omega)$, $\omega \in \Omega$ (see [Ch 1] for details). Ω is usually taken to be a function space corresponding to the set of all possible initial values of the velocity field $\underline{u}$ and in this situation the measure λ then describes the statistics of turbulent flows. Of particular interest is the possibility that a physically interesting λ could be constructed which is invariant under the flow determined by the Navier-Stokes equations. This is discussed in [Fo 1] where it is conjectured (with partial results proved) that the support Σ of any such invariant Borel probability λ must be a compact set of finite dimension in Ω. Thus, the set $S-Y$ could be supposed to have small λ-measure; alternatively, one could guess that the state space S for the Fourier model corresponds closely with Σ (if not, the existence of a nontrivial Y as above is suspect).

The above remarks make it natural to predict that the system (24) contains invariant sets of positive measure. In particular, we conjecture that the system is not mixing nor even ergodic (see [Ha 1]). It would appear that this actually follows from hypothesizing the existence of observables (collections of phase functions) which tend towards characteristic values in turbulent flows. However, one should keep in mind the arguments of the preceding paragraph. In particular, statistical mechanical arguments could conceivably explain such phenomena even in the face of ergodicity. In the 2-D case where there is an extra quadratic constant of the motion this seems unlikely but in the 3-D case this objection cannot be raised. Therefore, the conjecture is stronger in the 2-D case.

Some comments are in order concerning methodology in testing a dynamical system for statistical properties such as ergodicity. Since the system in question cannot be solved explicitly, it is necessary to approximate the solution by an appropriate difference scheme. In effect, this replaces the measure-preserving flow F_t by a sequence of approximate measure-preserving transformations. One may wonder in principle whether or not this introduces errors regardless of how small the time step. Let us consider testing ergodicity by comparing Cèsaro sums formed from many initial points and phase functions with the results of integrations of the phase functions over S, using the ergodic theorem (see [Ha 1]). The effect of the approximation is to replace

29)
$$\lim_{T \to \infty} \frac{1}{T} \int_0^T g(F_t p)\, dt$$

by the sum

30)
$$\lim_{\substack{N \to \infty \\ T \to \infty}} \frac{1}{N} \sum_{j=0}^N g(F_{\Delta t}^j p)$$

where $F_{\Delta t}$ approximates the flow for the time step Δt and $T/\Delta t = N$. For a fixed T, it is clear that (30) is an approximating Riemann sum for the integral in (29). So, although such a method can never _prove_ anything, strong evidence can nevertheless be obtained by taking T large enough and Δt small enough.

Finally, we report some results which have bearing on the topics discussed above.

(I) [Ch 3]. The Fourier-mode model for Burger's equation (12) is considered in regard to computer tests for ergodicity and mixing. The models considered are all small or medium-scale (less than 20 modes). The evidence is very strong that the system is not ergodic.

(II) [Hald I]. This paper looks at some very small-scale truncations in the form (19)-(2) in the 2-D case and actually constructs several constants of the motion. Thus, these systems are not ergodic.

(III) [Lee 1,2]. In these papers, Lee argues that Hald's results are special to the truncations involved and that a more general truncation will be ergodic.

(IV) [BS 1]. Here, both numerical and computational evidence is presented to the effect that quadratic phase functions of the 2-D system have equal time and phase means. More general phase functions are not considered.

(V) [DZ 1]. Energy and vorticity spectra are studied for the 2-D system (19)-(20) using a rather large-scale truncation (64 x 64 modes retained). Two disjoint sets of initial data are found which evolve under the flow towards distinct spectral distributions. However, the so-called "ergodic boundary" between these two sets may well be just a boundary between valid and invalid numerical results. If this is not so, the results strongly support the claim that the flow is not ergodic. Also, the results of this paper are related to the negative-temperature states considered in the vortex model.

(VI) Finally, we point out a result of a different flavor.

Let $n \geq 4$ and let I_n = unit n-cube. Then $M(I_n) = \{T\colon I_n \to I_n \mid T$
is a measure-preserving homeomorphism\} with the metric $\rho(T,S) =$
$\max \{\max_{x \in I_n} |Tx - Sx| \, , \, \max_{x \in I_n} |T^{-1}x - S^{-1}x|\}$ is a complete metric
space and the ergodic transformations from a residual set in the sense
of Baire category. See [MPV 1] for the proof. This result can be
extended to manifolds and so it has a direct bearing on the problems
considered here.

Bibliography

Basdevant, C. and Sadourney, R.: [1] Ergodic Properties of inviscid truncated models of two-dimensional incompressible flows, JFM 69, part 4 (1975).

Batchelor, G. K.: [1] The Theory of Homogeneous Turbulence, Cambridge Univ. Press (1960).

Bowen, R.: [1] Lecture 8 in the turbulence notes.

Chorin, A. J.: [1] Lectures on Turbulence Theory, Publish or Perish (1976).

[2] Computational Aspects of the Turbulence Problem, Proc. 2nd. Int. Conf. Num. Meth. Fluid Mechanics, Springer (1970).

[3] Numerical Experiments with a Truncated Spectral Representation of a Random Flow, unpublished.

[4] Numerical Study of slightly viscous flow, JFM 57 , part 4 (1973).

Deem, G. S. & Zabusky, N. J.: [1] Ergodic Boundary in Numerical Simulations of Two-dimensional Turbulence, Phys. Rev. Lett. 27 , no. 7 (1971).

Ebin, D. G & Marsden, J.: [1] Groups of diffeomorphisms and the motion of an incompressible fluid, Ann. of Math. 92 (1970).

Foias, C.: [1] Ergodic Problems in functional spaces related to the Navier-Stokes equations, Proc. Int. Conf. Funct. Annal. rel. Topics, Tokyo (1969).

Fox, D. G. & Orszag, S. A.: [1] Inviscid dynamics of two-dimensional turbulence, The Physics of Fluids 16, no. 2 (1973).

Gollub, J. P. & Swinney, H. L.: [1] Onset of Turbulence in a Rotating Fluid, Phys. Rev. Letters 35, no. 14 (1975).

Hald, O.: [1] Constants of motion in models of two-dimensional turbulence, The Physics of Fluids 19, no. 6 (1976).

Halmos, R. P.: [1] Lectures on Ergodic Theory, Chelsea Publishing Co. (1956).

Hopf, E.: [1] Statistical hydromechanics and functional calculus, J. Rat. Mech. Anal. 1 (1952).

Khinchin, A. I.: [1] Mathematical Foundations of Statistical Mechanics, Dover (1949).

Lee, J.: [1] How Many isolating Constants of Motion in 2-D Turbulence, preprint.

[2] Isolating Constants of Motion for the Homogeneous Turbulence of Two and Three Dimensions, J. of Math. Phys. 16, no. 7 (1975).

Lo, R. K. C. & Ting, L.: [1] Studies of the merging of vortices, Phys. of Fluids 19, no. 6 (1976).

Montgomery, D.: [1] Two-dimensional vortex motion and "negative temperatures," Phys. Lett. 39A, no. 1 (1972).

Montgomery, D. & Joyce G.: [1] Statistical mechanics of "negative temperature" states, preprint, Dept. of Physics and Astronomy, The Univ. of Iowa (1973).

Moser, J.: [1] Stable and Random Motions in Dynamical Systems, no. 77, Annals of Math. Stud., Princeton Univ. Press (1973).

Moser, J., Phillips, E. & Varadhan, S.: [1] Ergodic Theory: A Seminar, Courant Inst. Lecture Notes (1975).

Onsager, L.: [1] Statistical Hydrodynamics, Nuovo Cimento Supp. Al. Vol. VI, Serie IX, (1949).

Orszag, S. A.: [1] Analytical Theories of Turbulence, JFM $\underline{41}$, part 2 (1970).

[2] Numerical Simulation of Incompressible Flows Within Simple Boundaries. I. Galerkin (Spectral) Representations, Stud. in Appl. Math. $\underline{L}$, no. 4 (1971).

Seyler, C. E., Jr.: [1] Thermodynamics of two-dimensional plasmas or discrete line vortex fluids, The Phys. of Fluids $\underline{19}$, no. 9 (1976).

Vol. 460: O. Loos, Jordan Pairs. XVI, 218 pages. 1975.

Vol. 461: Computational Mechanics. Proceedings 1974. Edited by J. T. Oden. VII, 328 pages. 1975.

Vol. 462: P. Gérardin, Construction de Séries Discrètes p-adiques. »Sur les séries discrètes non ramifiées des groupes réductifs déployés p-adiques«. III, 180 pages. 1975.

Vol. 463: H.-H. Kuo, Gaussian Measures in Banach Spaces. VI, 224 pages. 1975.

Vol. 464: C. Rockland, Hypoellipticity and Eigenvalue Asymptotics. III, 171 pages. 1975.

Vol. 465: Séminaire de Probabilités IX. Proceedings 1973/74. Edité par P. A. Meyer. IV, 589 pages. 1975.

Vol. 466: Non-Commutative Harmonic Analysis. Proceedings 1974. Edited by J. Carmona, J. Dixmier and M. Vergne. VI, 231 pages. 1975.

Vol. 467: M. R. Essén, The Cos $\pi\lambda$ Theorem. With a paper by Christer Borell. VII, 112 pages. 1975.

Vol. 468: Dynamical Systems – Warwick 1974. Proceedings 1973/74. Edited by A. Manning. X, 405 pages. 1975.

Vol. 469: E. Binz, Continuous Convergence on C(X). IX, 140 pages. 1975.

Vol. 470: R. Bowen, Equilibrium States and the Ergodic Theory of Anosov Diffeomorphisms. III, 108 pages. 1975.

Vol. 471: R. S. Hamilton, Harmonic Maps of Manifolds with Boundary. III, 168 pages. 1975.

Vol. 472: Probability-Winter School. Proceedings 1975. Edited by Z. Ciesielski, K. Urbanik, and W. A. Woyczyński. VI, 283 pages. 1975.

Vol. 473: D. Burghelea, R. Lashof, and M. Rothenberg, Groups of Automorphisms of Manifolds. (with an appendix by E. Pedersen) VII, 156 pages. 1975.

Vol. 474: Séminaire Pierre Lelong (Analyse) Année 1973/74. Edité par P. Lelong. VI, 182 pages. 1975.

Vol. 475: Répartition Modulo 1. Actes du Colloque de Marseille-Luminy, 4 au 7 Juin 1974. Edité par G. Rauzy. V, 258 pages. 1975. 1975.

Vol. 476: Modular Functions of One Variable IV. Proceedings 1972. Edited by B. J. Birch and W. Kuyk. V, 151 pages. 1975.

Vol. 477: Optimization and Optimal Control. Proceedings 1974. Edited by R. Bulirsch, W. Oettli, and J. Stoer. VII, 294 pages. 1975.

Vol. 478: G. Schober, Univalent Functions – Selected Topics. V, 200 pages. 1975.

Vol. 479: S. D. Fisher and J. W. Jerome, Minimum Norm Extremals in Function Spaces. With Applications to Classical and Modern Analysis. VIII, 209 pages. 1975.

Vol. 480: X. M. Fernique, J. P. Conze et J. Gani, Ecole d'Eté de Probabilités de Saint-Flour IV–1974. Edité par P.-L. Hennequin. XI, 293 pages. 1975.

Vol. 481: M. de Guzmán, Differentiation of Integrals in R^n. XII, 226 pages. 1975.

Vol. 482: Fonctions de Plusieurs Variables Complexes II. Séminaire François Norguet 1974–1975. IX, 367 pages. 1975.

Vol. 483: R. D. M. Accola, Riemann Surfaces, Theta Functions, and Abelian Automorphisms Groups. III, 105 pages. 1975.

Vol. 484: Differential Topology and Geometry. Proceedings 1974. Edited by G. P. Joubert, R. P. Moussu, and R. H. Roussarie. IX, 287 pages. 1975.

Vol. 485: J. Diestel, Geometry of Banach Spaces – Selected Topics. XI, 282 pages. 1975.

Vol. 486: S. Stratila and D. Voiculescu, Representations of AF-Algebras and of the Group U (∞). IX, 169 pages. 1975.

Vol. 487: H. M. Reimann und T. Rychener, Funktionen beschränkter mittlerer Oszillation. VI, 141 Seiten. 1975.

Vol. 488: Representations of Algebras, Ottawa 1974. Proceedings 1974. Edited by V. Dlab and P. Gabriel. XII, 378 pages. 1975.

Vol. 489: J. Bair and R. Fourneau, Etude Géométrique des Espaces Vectoriels. Une Introduction. VII, 185 pages. 1975.

Vol. 490: The Geometry of Metric and Linear Spaces. Proceedings 1974. Edited by L. M. Kelly. X, 244 pages. 1975.

Vol. 491: K. A. Broughan, Invariants for Real-Generated Uniform Topological and Algebraic Categories. X, 197 pages. 1975.

Vol. 492: Infinitary Logic: In Memoriam Carol Karp. Edited by D. W. Kueker. VI, 206 pages. 1975.

Vol. 493: F. W. Kamber and P. Tondeur, Foliated Bundles and Characteristic Classes. XIII, 208 pages. 1975.

Vol. 494: A Cornea and G. Licea. Order and Potential Resolvent Families of Kernels. IV, 154 pages. 1975.

Vol. 495: A. Kerber, Representations of Permutation Groups II. V, 175 pages. 1975.

Vol. 496: L. H. Hodgkin and V. P. Snaith, Topics in K-Theory. Two Independent Contributions. III, 294 pages. 1975.

Vol. 497: Analyse Harmonique sur les Groupes de Lie. Proceedings 1973–75. Edité par P. Eymard et al. VI, 710 pages. 1975.

Vol. 498: Model Theory and Algebra. A Memorial Tribute to Abraham Robinson. Edited by D. H. Saracino and V. B. Weispfenning. X, 463 pages. 1975.

Vol. 499: Logic Conference, Kiel 1974. Proceedings. Edited by G. H. Müller, A. Oberschelp, and K. Potthoff. V, 651 pages 1975.

Vol. 500: Proof Theory Symposion, Kiel 1974. Proceedings. Edited by J. Diller and G. H. Müller. VIII, 383 pages. 1975.

Vol. 501: Spline Functions, Karlsruhe 1975. Proceedings. Edited by K. Böhmer, G. Meinardus, and W. Schempp. VI, 421 pages. 1976.

Vol. 502: János Galambos, Representations of Real Numbers by Infinite Series. VI, 146 pages. 1976.

Vol. 503: Applications of Methods of Functional Analysis to Problems in Mechanics. Proceedings 1975. Edited by P. Germain and B. Nayroles. XIX, 531 pages. 1976.

Vol. 504: S. Lang and H. F. Trotter, Frobenius Distributions in GL_2-Extensions. III, 274 pages. 1976.

Vol. 505: Advances in Complex Function Theory. Proceedings 1973/74. Edited by W. E. Kirwan and L. Zalcman. VIII, 203 pages. 1976.

Vol. 506: Numerical Analysis, Dundee 1975. Proceedings. Edited by G. A. Watson. X, 201 pages. 1976.

Vol. 507: M. C. Reed, Abstract Non-Linear Wave Equations. VI, 128 pages. 1976.

Vol. 508: E. Seneta, Regularly Varying Functions. V, 112 pages. 1976.

Vol. 509: D. E. Blair, Contact Manifolds in Riemannian Geometry. VI, 146 pages. 1976.

Vol. 510: V. Poènaru, Singularités C^∞ en Présence de Symétrie. V, 174 pages. 1976.

Vol. 511: Séminaire de Probabilités X. Proceedings 1974/75. Edité par P. A. Meyer. VI, 593 pages. 1976.

Vol. 512: Spaces of Analytic Functions, Kristiansand, Norway 1975. Proceedings. Edited by O. B. Bekken, B. K. Øksendal, and A. Stray. VIII, 204 pages. 1976.

Vol. 513: R. B. Warfield, Jr. Nilpotent Groups. VIII, 115 pages. 1976.

Vol. 514: Séminaire Bourbaki vol. 1974/75. Exposés 453 – 470. IV, 276 pages. 1976.

Vol. 515: Bäcklund Transformations. Nashville, Tennessee 1974. Proceedings. Edited by R. M. Miura. VIII, 295 pages. 1976.

Vol. 516: M. L. Silverstein, Boundary Theory for Symmetric Markov Processes. XVI, 314 pages. 1976.

Vol. 517: S. Glasner, Proximal Flows. VIII, 153 pages. 1976.

Vol. 518: Séminaire de Théorie du Potentiel, Proceedings Paris 1972–1974. Edité par F. Hirsch et G. Mokobodzki. VI, 275 pages. 1976.

Vol. 519: J. Schmets, Espaces de Fonctions Continues. XII, 150 pages. 1976.

Vol. 520: R. H. Farrell, Techniques of Multivariate Calculation. X, 337 pages. 1976.